AF560751

UNDERSTANDING BIOLOGICAL AGING

UNDERSTANDING BIOLOGICAL AGING

By

Dr. G.P. Garg

&

Dr. M.Prakash

DISCOVERY PUBLISHING HOUSE PVT. LTD.

NEW DELHI-110 002

First Published-2010

ISBN 978-81-8356-517-2

Published by:

DISCOVERY PUBLISHING HOUSE PVT. LTD.

4831/24, Ansari Road, Prahlad Street,
Darya Ganj, New Delhi-110002 (India)
Phone: 23279245 • Fax: 91-11-23253475
E-mail: parul.wasan@gmail.com
info@discoverypublishinggroup.com
Website: www.discoverypublishinggroup.com

Printed at:

Sachin Printers
Delhi

Preface

The present title "Understanding Biological Aging" has been written for those students interested in careers in diverse fields of biological sciences. It provides a structured approach to learning by covering all the important topics in a uniform, systematic format. The book has been comprehensively designed incorporating recent advances in this fast moving field. It also provides accessible information on biological aging in compact form for undergraduate students in biology and related life sciences. It is intelligible to the educated layman, though it deals with some complex ideas. It is an adequate text for all the requirements of students in this area. In addition, busy lecturers who require a quick reference compendium will find it useful, particularly for tutional planning. Simple, yet hopefully clear figures and tables are provided throughout the book.

The over-riding goal of this book, and indeed of the whole *Understanding series*, is to present the essential information concering biological aging in a compact, readily accessible form which leads itself to student learning and revision. The convergence of various approaches has generated a rich panorama of detail, the significance of which we are still attempting to unraval. The present text has been written as an introduction to this rapidly growing field.

To make the work more comprehensive and informative, the author has consulted many authoritative books, research journals, abstracts, monographs etc., so there can be no claim to originality except in the manner of treatment.

The author expresses his thanks to his friends and colleagues whose continue inspirations have initiated him to bring out this book.

The author expresses his gratitude to Mr. Wasan and staff of M/s Discovery Publishing House Pvt. Ltd. for their whole hearted co-operation in the publication of this book.

In the mean time, the author will remain sincerely responsible for any shortcomings of the book and be grateful to the readers for their suggestions and constructive criticism for the continuous betterment of the book. He takes this opportunity to appeal to the readers to send their suggestions straightaway to his Publisher.

Author

Contents

1

INTRODUCTION

In a now-classic experiment, Smith and Whitney showed heterogeneity in the doubling potential of individual cells from clonally derived populations, suggesting that stochastic factors determine the replicative potential of cells.

Other studies have shown that the fraction of cells able to divide decreases progressively with increasing population doubling, using BrdU labeling, Ki67 staining, and p53-reporter assay Data also show that the percentage of cells stained with γ-H2AX, which is a marker for senescence-associated DNA damage foci, formed in response to short, functionally uncapped telomeres, increases with population doubling.

The data are consistent in showing that a percentage of cells drop out of the cell cycle, and that this percentage increases with the age of the culture. The properties of these early "*drop-outs*" can be determined after physically separating cells in a mass culture.

Senescent cells differ from their proliferating counterparts by a wide variety of parameters, including gene expression pattern, cell and nuclear size, shape and structure, mitochondrial function and production of *reactive oxygen species* (ROS), accumulation of age pigment (lipofuscin), and activity of β-galactosidase at near-neutral pH (senescence-associated β-galactosidase, sen-β-gal), and others. Not all of these markers are readily suited for physical separation of cells by fluorescence-activated cell sorting (FACS).

More importantly, most potential markers differ only gradually between proliferating and senescent cells, and no single marker known so far is completely specific. Therefore, we describe here various methods by which senescent cells can be identified in a mass culture and physically sorted. Some of these methods use antibody staining of nuclear antigens and thus require permeabilization and fixation of cells, which severely limits the analytical choices after sorting.

However, these methods will very often produce clearly separated populations of senescent and nonsenescent cells. In contrast, there are few markers that allow sorting of live senescent cells as, for instance, cell size, lipofuscin content, or live fluorescent stains. These parameters, however, are generally homogeneously distributed between young and senescent cells, and cut-offs must be set arbitrarily in the sorting process. That means that quantitative estimates of frequencies of senescent vs nonsenescent cells cannot be obtained and sorting will result in an enrichment of senescent cells, but not in the production of a "*pure*" population.

This problem can be overcome by transfecting cells with *green fluorescent protein* (GFP)-expressing reporter constructs for genes transcriptionally regulated by senescence. We have used an EYFP reporter driven by the p21 promoter to FACS-sort senescent cells, and this tended to give a clearer separation of senescent cells than "*natural*" live cell markers.

Possible disadvantages with this approach are that random insertion of the reporter (e.g., by retroviral transfer) might by itself induce an additional level of heterogeneity, whereas targeted insertion followed by selection might leave only very little time to do the actual experiments in a senescing cell strain. However, the production of cells bearing stable expression reporters is outside the limited scope of this chapter and will not be reviewed here.

Materials

Cell Culture

1. Dulbecco's modified Eagle's medium (DMEM) with 10% fetal bovine serum, 1% penicillin/streptomycin, and 1% Glutamate.
2. Solution of trypsin-EDTA (5 g of trypsin, 2 g of EDTA per liter of distilled water).

Immunofluorescence (Nuclear Antigens Ki67 and γ-H2A.X)

1. PBG-Triton: (0.2% fish skin gelatin, 0.5% bovine serum albumin [BSA] in phosphate-buffered saline [PBS] and 0.5% Triton X-100).
2. Paraformaldehyde (2%).
3. Fluorescein isothiocyanate (FITC)-labeled monoclonal antibody AntiKi67.

4. Mouse monoclonal anti-γ-H2A.X.
5. Fluorescein-conjugated secondary antibody AlexaFluor 594.

BrdU Staining

1. Wash buffer: 1X Ca^{2+}- and Mg^{2+}-free PBS with 0.5% w/v bovine serum albumin BSA (Sigma).
2. Denaturing solution: 2 *M* HCl in 1X Ca^{2+}- and Mg^{2+}-free PBS with 0.5% w/v BSA (Sigma) (dissolve BSA in PBS prior to adding HCl to avoid permanent precipitation of the BSA).
3. Dilution buffer: 1X Ca^{2+}- and Mg^{2+}-free PBS with 0.5% v/v Tween-20 (Sigma), 0.5% w/v BSA (Sigma).
4. FITC-labeled AntiBrdU monoclonal antibody or FITC-labeled immunoglobulin (Ig)G control. *Note:* BrdU is a known carcinogen.
5. Propidium iodide (PI) (*Note*: PI is known to be *toxic* and *carcinogenic*).
6. Ice-cold 70% ethanol.
7. 0.1 *M* sodium borate (Sigma), pH 8.5.

FACS Analysis and Sorting

1. FACS calibration beads of different sizes and/or fluorescence intensities or Partec 3 *μM* calibration beads.
2. 1% Triton-X-100, degassed PBS 1X.

Vital Staining for ROS and MMP

1. Dihydrorhodamine 123 (DHR123) is available from Molecular Probes either as powder (10 mg vial) or as ready-to-use 5 m*M* stock solution in dimethylsulfoxide (DMSO). A 10 m*M* stock solution of DHR123 is prepared by adding 2.89 mL DMSO to the 10-mg vial. As DHR123 is readily oxidized by air, the stock solution is immediately aliquoted under N_2 atmosphere in 100 μL each and frozen at –20°C.
2. 5,5',6,6'-tetrachloro-1,1',3,3' - tetraethylbenzimidazolylcarbocyanine iodide (JC-1) (5 mg). To prepare a stock solution, dilute JC-1 at 2.5 mg/mL in DMSO by adding 2 mL of DMSO to the 5 mg vial. To avoid repeated freeze/thaw cycles, aliquot the stock immediately into portions, each sufficient for 1 d of experimental work, and store them at –20°C until required for use. Staining procedure must be carried out under no direct, intense light and incubation in the dark, because JC-1 is light sensitive.
3. RPMI 1640 phenol-free without additives.
4. DMEM without additives.

Methods

Setting Up the Flow Cytometer

FACS sorters vary considerably in complexity. There are highly sophisticated closed-circuit, multichannel, high-speed sorting machines operated by highly trained specialist personnel and dedicated exclusively to sorting. However, sorting can be done sufficiently well on much simpler (and cheaper) machines that are typically characterized by a sorting chamber fitted to an existing flow cytometer. On these machines, sorting will be the exception rather than the rule. In these settings especially, a number of issues are well worth keeping in mind when setting out to perform a cell-sorting experiment.

1. Preparation and calibration of the flow cytometer: sheath fluid must be of ultra-pure quality and should be de-gassed before use, especially for sorters that rely on piezo crystal action. The system is calibrated using fluorescent beads to ensure optimum performance and reproducibility. Sorting efficiency and the quality of separation are tested using a mixture of two preparations of calibration beads of different sizes. Re-analysis of the sorted fractions should show more than 20-fold enrichment of the sorted fraction. Re-analysis of a defined volume can also be used to count the number of sorted particles obtained and thus to optimize the sorting conditions.
2. Sterility: sterile sorting is possible even in simple flow cytometers. We sterilize the system by running 0.1% hypochlorite solution both as sample and sheath fluid for 20-30 min and thoroughly swab external orifices with the same solution. Following bleaching, sterile, de-gassed 1X Ca^{2+}- and Mg^{2+}-free PBS was run through for 10 min in order to remove residual bleach, which could have a cytotoxic effect. For the sorting experiments, sheath fluid is autoclaved and the container for sheath fluid is replaced by an autoclavable glass bottle. FACS tubes are presterilized with 0.1% bleach neutralized by washing with 1X Ca^{2+}- and Mg^{2+}-free PBS. Cells are sorted directly into flasks with DMEM medium containing 5% penicillin/streptomycin in order to minimize the possibility of infection.
3. Sorting speed: senescent fibroblasts are large and sensitive to shear stress. Although some systems offer extremely high sorting speeds, we find that the quality and integrity of the recovered cells decreases at higher speeds. We use rates of about 300 cells/s as a

safe upper margin to sort cells for analytical purposes. To obtain live senescent cells for further cultivation, we recommend sorting speeds of not more than 100 cells/s.

Sorting of Fixed Cells

Following antibody staining for nuclear antigens

A number of nuclear antigens are characteristic of either cycling or noncycling cells, respectively. Ki67 and PCNA are expressed in cycling cells irrespective of the cell cycle phase (G_1, S, G_2, and M phases) but are absent in resting cells (G_0), and thus are possible "*negative*" markers for senescent cells. It has been shown that the fraction of Ki67-positive nuclei declines continuously with the population-doubling level in human fibroblasts and mesothelial cells. The inhibitors of cyclin-dependent kinases p16 and p21 are upregulated in senescence and can thus be used as "*positive*" markers for senescent cells. It is not clear yet whether p16 and p21 are upregulated alternatively in different cells or sequentially.

The persistent formation of DNA damage foci and, more specifically, the phosphorylation of Ser-139 of histone H2A.X (γ-H2A.X formation) by ATM/ATR kinases has recently been shown to be another marker for senescence induced by uncapped telomeres or unrepaired DNA damage. The frequency of cells positive for γ-H2A.X increases steeply as cultures approach replicative senescence. Reliable antibodies for all of these markers are available from a number of major distributors.

1. Cells are incubated in 1 mL of 2% paraformaldehyde in PBS for 10 min at room temperature.
2. Paraformaldehyde is removed and cells are washed twice with PBS.
3. To permeabilize: cells are incubated for 45 min at room temperature with 1 mL PBG-Triton.
4. Primary antibody at a 1/2000 dilution (diluted in PBG-Triton) is added. Cells are incubated for 1 h at room temperature with gentle agitation.
5. Cells are washed twice with PBG-Triton.
6. Cells are incubated for 45 min to 1 h with fluorescein-conjugated secondary antibody (1/4000) diluted in PBG-Triton.
7. After washing twice with 1X PBS, cells are resuspended in 1X PBS at a density of about 10^6/mL and analyzed by FACS with 488 nm excitation. The cell population is defined in a forward

(FSC) vs sideward scatter (SSC) dotplot. Apoptotic cells (recognized by lower FSC and higher SSC) and debris (low in both FSC and SSC) are gated out. FL1 (or any other fluorescence channel if a different fluorophore is used) is set at logarithmic. Using isotype controls, the gain for FL1 is set to obtain unstained cells within the first decade. The sorting gate is set at the minimum between unstained and stained cells, and sorting is started at a rate of about 300 cells/s.

Sorting of fixed cells after BrdU incorporation

The absence of DNA replication for prolonged time spans as shown by the failure to incorporate nucleotide analogs such as BrdU is regarded as one of the best criteria of cellular senescence. On a cell population level, this marker appears to be more specific than, for instance, staining for senescence-as sociated β-galactosidase. However, BrdU incorporation (or rather its absence) is less specific for positive selection of senescent cells, as this fraction might contain a significant number of cells that are only temporarily withdrawn from the cell cycle. The technique involves DNA denaturation at low pH and leads to significant DNA degradation.

1. Growing cells are treated for 48 h with 20 μM BrdU. Cell density is adjusted such that the culture is still in exponential growth phase at the end of this period. Cells are harvested and counted.
2. Cells are centrifuged at 300*g* for 5 min, washed in 1X Ca^{2+}- and Mg^{2+}-free PBS, and re-centrifuged for 5 min.
3. The supernatant is aspirated and the pellet is loosened by tapping the tube. Ice-cold 70% ethanol is added dropwise while vortexing to a final concentration of 100 $\mu L/10^6$ cells and cells are fixed by incubation at room temperature for 20 min.
4. Cells are resuspended in 1 mL wash buffer and centrifuged at 300*g* for 5 min. The supernatant is aspirated and the pellet is resuspended in fresh denaturing solution at a concentration of 150 $\mu L/10^6$ cells.
5. Cells are denatured by incubation at room temperature for 20 min.
6. Cells are resuspended in 1 mL wash buffer and centrifuged at 300*g* for 5 min.
7. The supernatant is discarded and the pellet is resuspended in 0.5 mL 0.1 *M* sodium borate, pH 8.5, and incubated at room temperature for 2 min to neutralize residual acid. One milliliter

wash buffer is added and the cell sample volume is divided into tubes for test and control. These are centrifuged at 300*g* for 5 min and the supernatant is aspirated.

8. FITC-labeled AntiBrdU monoclonal antibody or FITC-labeled IgG control are diluted in dilution buffer with 20 μL of antibody per 30 μL of buffer.
9. Cells are resuspended in diluted antibody at a concentration of 50 $\mu L/10^6$ cells and incubated in the dark at room temperature for 30 min. One milliliter wash buffer is added and cells are centrifuged at 400*g* for 5 min.
10. The supernatant is aspirated. At this step, cells can optionally be stained with PI to reveal the cell cycle distribution at the end of the BrdU incubation period. Stained or unstained cells are diluted with 1 mL 1X Ca^{2+}- and Mg^{2+}-free PBS for analysis.
11. Cells are analyzed with 488 nm excitation. The cell population is defined in an FSC vs SSC dotplot. BrdU incorporation is measured in FL1, set at logarithmic. Using isotype controls, the gain for FL1 is set to obtain unstained cells within the first decade. The sorting gate is set at the minimum between unstained and stained cells. Alternatively, if cells are PI-stained, a polygon gate can be used to define sorted cells in a FL1/Fl3 dotplot. Sorting is performed at a rate of about 300 cells/s.

Sorting of Live Cells

Sorting of live cells is ideal insofar as it poses no limits whatsoever for successive analysis. However, there are no binary markers for senescence in genetically unmodified cells. This means that the choice of the sorting gate position and the resulting enrichment of senescent cells are, to some extent, arbitrary.

Size and autofluorescence

Increased cell size has long been known as a characteristic of senescent cells. Size, as measured by FSC, has been used to FACS-sort populations of cells with a senescent phenotype from a culture of immortalized fibroblasts. The cellular content of lipofuscin, the undegradable, autofluorescent "*age pigment*," increases with time not only in postmitotic cells, but also in proliferating ones. This increase becomes faster as cells approach senescence, probably both because of higher oxidative stress resulting in faster accumulation and slower '*dissolution*' onto daughter cells due to slower cycling. Combining size and autofluorescence (measured in FL-1) as sort parameters has allowed good separation of live senescent and proliferating cells, as confirmed by using various markers of senescent cells including morphology and growth rates, senescence-associated-β-gal, and γ-H2A.X staining and BrdU incorporation.

1. Cells in exponential growth phase are trypsinized, collected in DMEM plus 10% serum at 4°C, and immediately used for analysis and sorting in a FACS sorter.
2. Autofluorescence of unfixed cells is measured in FL1 (logarithmic). Cell size is monitored by FSC. FSC can be measured using either linear or logarithmic scale. Using linear scale for FSC, better resolution can be achieved.
3. The population of live cells is defined in a FSC/SSC dotplot. Apoptotic cells and debris are gated out. Sorting gates are set in a FSC/FL1 dotplot. Using the upper and lower quartiles with respect to both FSC and FL1 to sort senescent and young proliferating cells, respectively, results in good resolution between the sorted populations and good enrichment of senescent cells.
4. Cells are sorted at a rate of 100–300 cells/s and collected on ice. If cells must be concentrated, low centrifugation speed (≤ 200*g*) should be used.

Live cell stains for ROS or MMP

Early- and late-passage fibroblasts differ in mitochondrial function and, consequently, in oxidative stress levels. Assessment of ROS levels in whole cells using fluorescent probes has shown that in vitro senescence, like in vivo aging, is associated with high levels of endogenous ROS. Recently, we have shown using the fluorescent probe JC-1 that senescent cells have lower membrane potential than proliferating ones. Using JC-1 staining, Santos and colleagues sorted cells with different mitochondrial membrane potential and analysed the sorted populations for mitochondrial DNA damage.

1. Cells in exponential growth phase are trypsinized, collected in DMEM plus 10% FCS, counted, washed once with PBS, and centrifuged for 5 min at 400*g*.
2. Supernatant is discarded.

DHR 123 staining of peroxides

1. Cell pellet is resuspended in 5 mL serum-free DMEM containing 15 μL DHR123 stock solution (final DHR123 concentration 30 μM).
2. Cells are incubated at 37°C for 30 min in the dark.

3. After incubation, cells are centrifuged (400*g*, 5 min) and supernatant discarded.
4. The pellet is resuspended in 3 mL of serum-free DMEM and is immediately used for analysis and sorting.
5. The population of live cells is defined in a FSC/SSC dotplot and apoptotic cells and debris are excluded from sorting. Sorting gates are defined in Fl1 (logarithmic), conveniently in a FSC/FL1 dotplot. Use the left and right quartiles with respect to FL1 to sort cells with highest (high FL1) and lowest (low FL1) peroxide levels.

JC-1 staining for determining MMP

1. 1×10^5 cells are resuspended in 500 μL of RPMI 1640 phenol-free without supplements (Sigma) and 1 μL of JC-1 stock solution is added to give a final concentration of 1 μg/mL.
2. The cells are incubated for 30 min at 37°C.
3. The cells are harvested by centrifugation at 300*g*, 4°C for 5 min, washed with ice-cold PBS, and resuspended in 200 μL PBS at 4°C.
4. The population of live cells is defined in a FSC/SSC dotplot and apoptotic cells and debris are excluded from sorting. FL1 and FL3 are set at logarithmic, and sorting gates are defined in a FL3/FL1 dotplot. The quartile of all cells with simultaneously highest FL1 and lowest FL3 will be enriched for cells with low membrane potential, i.e., senescent cells. Accordingly, the quartile of cells with the lowest FL1 and highest FL3 constitutes the cells with highest mitochondrial membrane potential.
5. To ensure that JC-1 staining is working, it is recommended that a control sample, in which mitochondria of all cells have been depolarised, be used. Treating cells with drugs able to collapse $\Delta\Psi$, such as the proton translocator carbonyl cyanide p-(trifluoromethoxy) phenylhydrazone (FCCP 20 μM, 30 min), results in a dramatic change of the fluorescence distribution.

Notes

1. Before re-analysis, thorough cleaning of the tubings in the flow cytometer is necessary to obtain reliable quantitative results. The cleaning function in most flow cytometers is not sufficient for this. We use 1 mL 1% Triton, followed by 2 mL distilled water, followed by 1 mL PBS.
2. Most of the nuclear markers used in this section are at least somewhat sensitive to reversible growth arrest, e.g., by confluency

or serum starvation. For instance, p21 is elevated in reversibly arrested G_0 cells. On the other hand, evidence suggests that PCNA can occasionally be expressed in noncycling G_0 cells. Background levels of DNA damage foci are slightly higher in G_1 than in G_1, possibly because of recognition of stalled replication forks, but might be suppressed in the absence of serum. Cells should, therefore, be reproducibly subconfluent at the time of sampling for sorting.

3. Permeabilization conditions (type of detergent, length of incubation time) must be adjusted according to the cells in question. Permeabilization must be gentler than for immunohistochemistry in order to keep cells intact while passing them through the flow cytometer. Ideally, the cells should still appear as a well defined population in FSC vs SSC; at the same time, the permeabilization should allow efficient antibody penetration.
4. Antibody dilutions are given as a starting example only; concentrations must be titrated out for every combination of cell strain and antibody. The indicated concentrations work well in MRC-5 fibroblasts, for mouse monoclonal γ-H2A.X, and fluorescein-conjugated secondary antibody.
5. We isolated DNA from MRC-5 fibroblasts after processing for BrdU staining and analyzed it by pulsed-field gel electrophoresis. Only heavily degraded DNA could be obtained, excluding all posterior analysis relying on high-molecular-weight DNA.
6. To obtain a more efficient selection of senescent cells, the BrdU incorporation time can be extended for up to 72 h. However, long-term BrdU treatment is cytotoxic. Care must be taken not to expose cells to light while they are incorporating BrdU, as this amplifies DNA strand break generation.
7. For staining of DNA, incubate the cells in PI (Sigma) at 10 μg/mL in PBS for 30 min in the dark at room temperature. Use 0.5 mL PI solution for 10^6 cells.
8. In growth-arrested cells, the rate of lipofuscin accumulation is increased because distribution of lipofuscin-containing lysosomes between daughter cells is the major if not the only way of diminishing lipofuscin levels.
9. If the instrument is carefully calibrated from run to run, gate positions defined in a standard population of cells can be used for different experiments to measure quantitative shifts in the ratio between senescent and nonsenescent cells.

10. Sorting times including storage of cells in PBS can be lengthy (a few hours), and sorting itself constitutes a stress to the cells. At the end of this process, human fibroblasts become highly sensitive to shear stress, resulting in the loss of the majority of cells during centrifugation at higher g force.
11. Inside live cells, the colorless dihydrorhodamine 123 is oxidized, e.g., by hydrogen peroxide in the presence of peroxidases, iron, or cytochrome c to form rhodamine 123. Moreover, DHR also reacts with peroxynitrite. The fluorescent product accumulates in mitochondria, even if the oxidation occurred in other organelles or the cytoplasm. Rhodamine 123 fluoresces both in FL1 (green) and FL3 (red), but we find detection to be more sensitive in FL1.
12. At room temperature and under ambient light, DHR123 fluorescence of human fibroblasts remains fairly constant for at least 1–2 h. If longer sorting times are necessary, cells should be protected from light and kept at 4°C.
13. The JC-1 probe exists as a green fluorescent monomer at low concentrations, but forms aggregates at higher concentrations that exhibit a broad excitation spectrum and an emission maximum at 590 nm. JC-1 can be excited at 488 nm and detected in bivariate mode using the green channel for the monomer and the red channel for the aggregate. Mitochondrial uptake of JC-1 is dependent on membrane potential, with higher uptake (and higher probability of red fluorescence) at higher potential. Thus, the FL3/FL1 ratio is an indicator of mitochondrial membrane potential.

Cellular Senescence

Normal human cells irreversibly arrest growth with a large and flat cell morphology after a limited number of cell divisions, or challenge by potentially oncogenic insults such as direct DNA damage or expression of certain oncogenes, in culture and possibly in vivo. This process is termed cellular senescence, and was first described by Hayflick and colleagues in cultured human fibroblasts. Cellular senescence is now recognized as an anti-proliferative response and tumor suppressor mechanism. In addition, the accumulation of senescent cells in aged tissues is also thought to contribute to age-related pathologies.

It is generally accepted that human fibroblasts and other cell types senesce after repeated cell division because they eventually acquire one or more short, dysfunctional telomeres. Recent evidence suggests that cells also undergo senescence as a result of nontelomeric signals, such as those delivered by certain oncogenes, strong mitogenic signals, direct DNA damage, and chromatin remodeling agents. Because senescent cells are thought to contribute to a subset of age-related pathologies, its abrogation can potentially help in the treatment of these pathologies. On the other hand, because the senescence response is a tumor suppressor mechanism, its induction by therapeutic agents in tumors can facilitate cancer treatment. Thus, the identification of senescent cells is important in studying the effects of both senescence-inducing and senescence-abrogating agents.

Senescent cells can be identified by their failure to undergo DNA synthesis under optimal culture conditions, or by genes that are differentially expressed during senescence. The assays for these characteristics are either nonspecific or tedious and time-consuming. For example, DNA synthesis measurements do not distinguish senescent cells from quiescent or terminally differentiated cells. Moreover, quiescent and terminally differentiated cells also show a down-regulation of proliferation-associated genes and upregulation of growth inhibitory genes. Several years ago, by serendipity we found that senescent cells expressed a β-galactosidase activity, which is histochemically detectable at pH 6.0.

We termed this activity the senescence-associated β-galactosidase, or SA-β-gal, and suggested it could be a good biomarker to identify senescent cells in culture and in in vivo. This marker was expressed by senescent, but not pre-senescent or quiescent, fibroblasts, nor by terminally differentiated keratinocytes. SA-β-gal also showed an age-dependent increase in dermal fibroblasts and epidermal keratinocytes in skin samples from human donors of different age. Although we showed that this marker was not a perfect senescence- or age-dependent marker (for example, it was also expressed when cells were maintained at confluence for prolonged periods), we showed that it was tightly associated with the senescent phenotype and increased in frequency in aged tissues, consistent with the accumulation of senescent cells with age in vivo.

Several subsequent studies have reinforced the idea that SA-β-gal is a useful biomarker for the detection of senescent cells in culture, as well as in vivo in rodents and primates.

Cellular senescence can be induced by multiple methods. The most common is continuous passaging of cells in culture. Other methods of senescence induction include:

1. DNA damage by radiation (X- or γ-rays, UV) or DNA-interacting drugs such as bleomycin.
2. Oxidative stress, for example from exposure to hyperoxia (40-50% oxygen), H_2O_2 or inhibition of reactive oxygen species (ROS)-scavenging enzymes, such as superoxide dismutase.
3. Oncogenic or hyperproliferative signals, for example by the expression of activated oncoproteins such as RAS or RAF or by mitogenic stimuli such as that caused by the overexpression of E2F1 or ETS2.
4. Overexpression of certain tumor suppressor proteins such as ARF, p16, or PML.

The induction of cellular senescence in most of the above-referenced cases was confirmed by failure of the cells to synthesize DNA and an increase in SA-β-gal staining and/or costaining for SA-β-gal and other recently described senescence markers, such as p16 tumor suppressor protein.

Materials

Cell culture

1. Dulbecco's modified Eagle's medium (DMEM) supplemented with 10% fetal bovine serum (HyClone).
2. 100X penicillin-streptomycin (10,000 U penicillin [base], 10 mg streptomycin [base]/mL utilizing penicillin G [sodium salt] and streptomycin sulfate in 0.85% saline).
3. Normal human fibroblasts: WI-38 and BJ strains (American Type Culture Collection [ATCC], Manassas,VA).

Fixation and SA-β-Gal staining of cultured cells

1. Phosphate-buffered saline (PBS); (10X stock): 1.37 *M* NaCl, 27 m*M* KCl, 100 m*M* Na_2HPO_4, 18 m*M* KH_2PO_4.
2. Fixing solution: 3.7% formaldehyde in PBS. Add 1 mL of 37% formaldehyde to 9 mL of PBS. The solution is freshly prepared for each experiment.
3. Staining solution: 1 mg/mL of X-gal, 40 m*M* citric acid/sodium phosphate buffer (pH 6.0), 5 m*M* potassium ferricyanide, 5 m*M* potassium ferrocyanide, 150 m*M* NaCl, and 2 m*M* $MgCl_2$.
   (a) X-gal solution: X-gal is dissolved at 20 mg/mL in dimethylformamide (DMF) in dark-colored or aluminum foil-wrapped glass vials or similar containers to protect from light. The solution can be stored at –20°C for a few days.

(b) Citric acid/sodium phosphate buffer (0.2 *M*, pH 6.0): mix 36.85 mL of 0.1 *M* citric acid solution with 63.15 mL of 0.2 *M* sodium phosphate (dibasic) solution. Verify that the pH is 6.0. The buffer can be kept at room temperature for several months.

(c) Citric acid solution (0.1 *M*): citric acid monohydrate ($C_6H_8O_7 \cdot H_2O$) is dissolved at 0.1 *M* in water. The solution can be kept at room temperature for several months.

(d) Sodium phosphate solution (0.2 *M*): sodium dibasic phosphate (Na_2HPO_4) or sodium dibasic phosphate dehydrate ($Na_2HPO_4 \cdot H_2O$) is dissolved in water at 0.2 *M*. The solution can be stored at room temperature for several months.

(e) Potassium ferricyanide solution (100 m*M*): potassium ferricyanide is dissolved in water at 100 m*M* and stored at 4°C in a tube covered with aluminum foil to protect from light and can be stored for several months.

(f) Potassium ferrocyanide solution (100 m*M*): potassium ferrocyanide is dissolved in water at 100 m*M* and stored at 4°C in a tube covered with aluminum foil to protect from light and can be stored for several months.

Fixation and SA-β-Gal staining for tissue samples

1. Fixing solution: 1% formaldehyde in PBS.
2. Staining solution: the staining solution is same for cultured cells and tissue samples (described previously).
3. Counter staining solution: eosin (Sigma).

[³H]Thymidine labeling and autoradiography

1. [Methyl-^{3}H]thymidine (70–95 Ci/mmol, 1 mCi/mL).
2. PBS containing 100 mg/L $CaCl_2$ and 100 mg/L $MgCl_2$.
3. Methanol.
4. Photographic emulsion [NTB2, cat. no. 165 4433(^{3}H), Kodak, Rochester, NY]: the emulsion is diluted 1:2 or 1:3 with distilled water, aliquoted into plastic vials covered with aluminum foil to protect from light, and stored at 4°C.
5. Kodak D-19 Developer (Kodak).
6. Kodak Rapid-Fix (Kodak).
7. Giemsa working solution: dilute 1 mL of Giemsa stock solution (0.4 w/v in buffered methanol solution, pH 6.9, Fischer Scientific, Pittsburgh, PA) to total volume of 10 mL with phosphate buffer pH 6.0 (74 m*M* NaH_2PO_4, 9 m*M* Na_2HPO_4).

Immunostaining

1. Lab-Tek II Chamber Slide II.
2. Fixing solution: 3.7% formaldehyde.
3. Permeabilizing solution: 0.5% Triton in PBS.
4. Blocking solution: 0.5% BSA in PBS.
5. Antibody dilution buffer: 0.5% BSA in PBS.
6. Secondary antibody.
7. Mounting medium: VectaShield containing DAPI (4',6'-diamidino-2-phenylindole; Vector Laboratories, Burlingame, CA).

Methods

SA-β-Gal staining for adherent cultured cells

1. Plate 2-5 × 10^4 cells in 35-mm dishes or similar vessel and culture for 1-3 d.
2. Wash cells twice with PBS.
3. Fix cells with freshly prepared 3.7% formaldehyde in PBS for 3-5 min at room temperature.
4. Wash cells twice with PBS.
5. Add X-gal staining solution (1–2 mL per 35-mm dish).
6. Incubate cells with staining solution at 37°C (not in a CO_2 incubator).
7. Blue color is detectable in some cells within 2 h, but staining is generally maximal in 12-16 h. An example of SA-β-gal staining of presenescent and senescent WI-38 fibrobalst, and control and ARF-overexpres sing U2OS cells.

SA-β-Gal staining for tissue samples

1. Obtain biopsy specimens and rinse briefly in PBS to remove any blood.
2. Place in OCT compound in a Tissue-Tek Cryomold and flash-freeze in liquid nitrogen containing 2-methyl-butane.
3. Unused samples can be stored at -80°C, but the enzyme is not stable for very long after freezing. In general, samples should be processed immediately or within a few hours after freezing.
4. Cut 4-μm sections of the samples.
5. Place sections onto slides that have been treated with silane to make them adhesive.
6. Fix sections in 1% formalin in PBS for 1 min at room temperature.
7. Wash with PBS three times.

8. Immerse sections in SA-β-gal staining solution overnight.
9. Counterstain with eosin.
10. View by bright-field microscopy.

Thymidine-labeling for cultured cells

1. Label cells with 10 μCi/mL [methyl-^{3}H]thymidine for 3 d or more.
2. Warm the Kodak emulsion in 37°C water bath.
3. Rinse culture plates three times with Ca^{2+}- and Mg^{2+}-containing PBS.
4. Rinse plates three times with methanol.
5. Allow plates to air-dry for 10 min.
6. Add emulsion to the plates using a disposable transfer pipet in a dark room equipped with the safelight (Kodak filter No. 2, cat. no. 152 1525; 15-V light bulb [no less than 4 feet from the emulsion]). Remove excess emulsion from the plates using a pipet, such that the plates are covered by a thin layer of emulsion. Excess emulsion can be returned to the original vials. Avoid getting air bubbles in the emulsion.
7. Store plates in a light-tight container (generally covered with foil) for at least 18 h.
8. Add developer in the dark room with a transfer pipet and wait 3 min.
9. Wash plates twice with water.
10. Add fixer in the dark room and wait 5 min.
11. Bring plates from the dark room and wash several times with water.
12. Allow plates to air-dry for 10 min.
13. Add freshly made Giemsa working solution and wait 5 min.
14. Wash plates with water.
15. Determine the percent radiolabeled nuclei (%LN) by counting the number of total (blue plus black) and labeled (black) nuclei in several randomly chosen fields (generally 100–500 total nuclei). %LN = [labeled nuclei/total nuclei]/× 100.

SA-β-Gal staining with thymidine labeling for cultured cells

1. Label cells with 10 μCi/mL of [methyl-3H]thymidine for 3 d or more.
2. Wash, fix, and stain for SA-β-gal activity as described above.
3. After blue color develops, wash, coat cells with emulsion, develop, and fix as described above, eliminating the Giemsa staining step.

SA-β-Gal staining with immunostaining for cultured cells

1. Culture cells in four-well chamber slides.
2. Wash, fix, and stain for SA-β-gal activity as described above.
3. Wash with PBS twice after blue color is developed.
4. Permeabilize cells with 0.5% of cold Triton X-100 in PBS at 4°C.
5. Block slides with 0.5% BSA in PBS for 20 min.
6. Incubate slides with a primary antibody in 0.5% BSA either for 2 h at room temperature or overnight at 4°C.
7. Wash three times with PBS, 10 min each.
8. Incubate slides with secondary antibody in 0.5% BSA for 1 h at room temperature.
9. Wash three times with PBS, 10 min each.
10. Mount cells in mounting medium containing DAPI.
11. Blue-colored staining for SA-β-gal activity is recognized well by bright-field microscopy and immunostaining is detected by epifluorescence.

Notes

1. For convenience, small aliquots of neutral buffered solution containing 10% formalin for SA-3-gal staining and immunostaining is available. The solution is stored at room temperature, and each small container can be used up to a month after opening. For some cells or tissues, freshly prepared 2% formaldehyde + 0.2% glutaraldehyde in PBS preserves cell morphology somewhat better. Twenty-five percent glutaraldehyde can be obtained in small aliquots from Sigma and stored at -20°C.
2. Some cell types, such as mouse fibroblasts or human epithelial cells, stain less intensely for SA-β-gal. The staining intensity can sometimes be improved by decreasing the pH slightly. Try several pH ranges from 5.0 to 6.0 to optimize the staining conditions, making sure to include positive and negative controls. Most, if not all, cells stain positive at pH 4.0 because of endogenous lysosomal β galactosidase activity and regardless of senescence status, so exercise caution when lowering the pH of the staining solution.
3. Ca^{2+}- and Mg^{2+}-containing PBS is used to reduce cell detachment from the culture dish during the staining procedure.
4. Co-staining for SA-β-gal and other proteins by immunostaining is antigen- and antibody-specific. If background immunostaining is

high, which is not uncommon with senescent cells, blocking with 10% nonfat milk in PBS or diluting the primary and/or secondary antibodies may help.

5. If cultured fibroblasts are confluent for long periods, density-induced SA-β-gal staining, independent of senescence, may occur. Such staining generally disappears after the confluent cells are replated.
6. Direct freezing in liquid nitrogen may fracture the specimen. Thus, it is best to place samples in OCT on top of dry ice. Avoid repeated freeze–thawing of the samples, which will affect morphology and destroy the SA-β-gal enzymatic activity.
7. Some tissue structures, such as hair follicles and the lumens of eccrine glands, show strong age-independent staining.
8. Three days of labeling is required to distinguish senescent and proliferating cells accurately.
9. The background of the photographic emulsion can increase somewhat when used on senescent cells. If the emulsion is used after the expiration date, check the background on a blank slide before use. Coating of plates with emulsion should be done quickly before the emulsion becomes viscous at room temperature. In addition, the emulsion layer should be thin to optimize visualization of labeled nuclei.
10. Giemsa staining should be kept to 10 min or less to avoid possible detachment of the emulsion from the culture dishes.

2

Cellular Senescence

Nearly 40 years ago, Hayflick and colleagues reported that primary mammalian cells have a finite life span in tissue culture, suggesting that there may be intrinsic mechanisms that regulate cell divisions. Cultivation of primary cells over many generations eventually resulted in a reproducible loss of proliferative potential that has been termed senescence. Therefore, senescence may reflect organismal aging. Recent work has revealed that senescence is a very heterogeneous process. Representing a complex phenotype, the term senescence has now been applied to any form of growth arrest of cultured cells that occurs either after some period of time or following the overexpression of selected oncogenes. Importantly, the analysis of the various aspects and types of senescence has increased our understanding of multiple cellular processes, such as aging and carcinogenesis. Because a variety of stimuli, many of which are potentially oncogenic, induce a senescent phenotype, cellular senescence is believed to prevent cellular proliferation and transformation. In this context, cellular senescence may represent a tumor-suppressive mechanism.

The first well documented illustration of cellular senescence was that of replicative senescence, which is the limited capacity to replicate and a defining characteristic of most normal cells. Replicative senescence was described initially in cultured human fibroblasts. Subsequently, many diverse types of cells in different species have been shown to undergo replicative senescence. The majority of these studies were performed using cells cultured ex vivo, but a limited number of studies support the idea that cells also undergo senescence in vivo. Replicative senescence is mediated by telomere shortening, whereby the senescence arrest occurs before the telomeres become short enough to compromise chromosomal integrity. It has long been appreciated that telomeres, the repetitive DNA that caps the ends of linear chromosomes, are essential structures that prevent chromosome fusion and genomic instability. In the absence of telomerase, which adds telomeric repeats to telomeres, each cell division is associated with the loss of 30–150 bp of telomeric DNA and critically short telomeres eventually induce normal cells to senesce. The salient feature is that cells undergo replicative senescence when they acquire one or more critically short telomeres. Moreover, telomere shortening and the replicative senescence of some human cells can be abrogated by ectopic expression of telomerase. Our increasing understanding of the role of telomere shortening in the replicative aging of cultured cells now permits an examination of one major mechanism of cellular senescence.

Apart from the induction of replicative senescence by critical telomere shortening, stimuli having little, if any, impact on telomeres have been shown to cause normal cells to undergo growth arrest and harbor a senescence phenotype. These stimuli include DNA damage, chromatin remodeling, and oncogenic or potent mitogenic signals. Normal cells undergo telomere-independent senescence in response to oncogenes, such as activated *ras* or *raf*, or supraphysiological mitogenic signals, such as overexpressed mitogen-activated protein kinase (MAPK) or the E2F1 transcription factor. Overstimulation of the Ras/Raf/MEK/MAPK pathway provokes premature senescence irrespective of telomere length. Nevertheless, premature senescence induced by activated *ras* and replicative senescence initiated by telomere shortening both share common signaling pathways and morphological features. In primary human cells, this premature senescence program leads to the activation of p53 and p16, therefore possibly functioning as a tumor-suppressor mechanism. Tumor suppressors that induce cellular senescence include the p16 cyclin-dependent kinase inhibitor (CDKI), the p14/ARF regulator of MDM2, and promyelocytic leukemia protein (PML). p14/ARF is a central regulator of p53 and critical contributor to the tumor-suppressor function of p53. Thus, p53 is essential for the senescence response to short telomeres, DNA damage, and oncogenes as well as mitogenic signals. Other conditions that have been linked to the activation of senescence programs include oxidative stress and DNA damage.

The senescent phenotype comprises more than an arrest of cell proliferation. However, the hallmark of cellular senescence is an essentially irreversible arrest of cell division. Senescent cells arrest growth in the G1 phase of the cell cycle, and cannot be stimulated to resume proliferation by physiological mitogens. Nevertheless, senescent cells can be rescued from their senescent phenotype by suppressing wild-type p53 and on the contrary immortalized cells can be forced into senescence by overexpressing, e.g., *ras* or p16. Coupled with this growth arrest, senescent cells show selected changes in morphology and metabolism, and derangements in differentiated functions. Some of the senescence-associated changes that are common to many different cell types include cellular enlargement, increased lysosome biogenesis, and expression of a β-galactosidase having a pH optimum of 6.0 (senescent-associated β-galactosidase [SA-β-gal]).

Cellular senescence can be induced through several different mechanisms. Many of these mechanisms can be experimentally recapitulated by the introduction of defined genetic elements. The telomere-dependent replicative senescence can either be studied in primary cells, which lack substantial telomerase activity and have a limited replicative potential, or by the utilization of different means to inhibit telomerase. Cellular senescence by oncogenic stimulation can be induced by, e.g., overexpression of *ras* or p16 by retroviral or adenovirally mediated gene transfer. This can be even achieved in previously immortalized cells. Regardless of the mechanism by which cellular senescence is induced, it can be described as an arrested state in which the cell remains viable but displays altered patterns of gene and protein expression.

Materials

Cell Cultures

To study cellular senescence, whether replicative or premature, several types of primary cells from different species can be used. As examples, we describe the tissue culture procedures for primary human fibroblasts, two types of primary human epithelial cells, and *mouse embryonic fibroblasts* (MEFs).

1. Human fibroblasts are cultured in Dulbecco's modified Eagle's medium (DMEM) supplemented with 10% fetal bovine serum (FBS) (Sigma) and 1% penicillin-streptomycin in 5% CO_2-20% O_2 at 37°C. Serum starvation is carried out with 0.1% FBS in DMEM for 120 h after trypsinization of serially passaged cultures. Serum restimulation is carried out with 10% FBS in DMEM. Alternatively, for better initial growth conditions, primary human

fibroblasts can be cultured in DMEM/F12 medium plus 15% iron-fortified newborn calf serum plus 10 ng/mL epidermal growth factor (EGF).

2. Primary human keratinocytes (e.g., from skin, esophagus, oral cavity) are grown at 37°C and 5% CO_2 with keratinocyte-SFM medium (KSFM), supplemented with 40 μg/mL bovine pituitary extract (BPE; Invitrogen), 1.0 ng/mL rEGF, 100 U/mL penicillin, 100 g/mL streptomycin (Sigma), and a final calcium-concentration of 0.4 m*M*. For better growth conditions, these cell populations can initially be expanded through one to four serial passages by cocultivating with mitomycin-treated Swiss 3T3J2 cells in flavin adenine dinucleotide (FAD) medium, consisting of DMEM/F12 (1:1, vol/vol) medium, 5% calf serum, 10 ng of EGF per milliliter, 0.4 μg of hydrocortisone per milliliter, 5 μg of insulin per milliliter, 10×10^{-10} *M* cholera toxin (CT), 2×10^{-11} *M* triiodothyronine, and 1.8×10^{-4} *M* adenine.

3. Human mammary epithelial cells (HMECs) can be either obtained or isolated from tissue specimens from reduction mammoplasties as described by Stampfer and are cultured in mammary epithelial growth medium (MEGM) at 37°C with 5% CO_2. MEGM is a serum-free medium composed of modified MCDB 170 basal medium with supplements (EGF, insulin, BPE, and hydrocortisone). Most of the normal HMEC available represent postselection cells beyond the so called mortality stage 0, which display long-term growth in MCDB 170. These cells are particularly useful in molecular and biochemical studies since they provide a virtually unlimited supply of uniform batches of normal human epithelial cells.

4. Primary MEFs derived from wild-type day 13.5 embryos are prepared as described. MEF cultures are maintained in DMEM (Invitrogen) supplemented with 10% FBS and 1% penicillin G/streptomycin. MEFs in vitro are adherent cells with good ability for proliferation. In room temperature, the digestive time of 0.25% trypsin should be about 3–5 min.

Buffers

1. SA-β-Gal staining solution consists of 1 mg/mL 5-bromo-4-chloro-3-indolyl β-galactoside (X-gal), 5 m*M* potassium ferrocyanide, 5 m*M* potassium ferricyanide, 1 m*M* $MgCl_2$ in phosphate-buffered saline (PBS) at pH 6.0.
2. Western blot RIPA buffer consists of 10 m*M* Tris-HCl (pH 7.4), 150 m*M* NaCl, 1% Nonidet P-40 (NP-40), 0.1% sodium dodecyl

sulfate (SDS), 0.1% DOC, 1 m*M* EDTA, 2 m*M* Na_3PO_4, 1 m*M* phenylmethylsulfonyl fluoride, and Complete Mini, mixture of protease inhibitors. Western blot TBST buffer consists of 10 m*M*Tris, 150 m*M*NaCl, pH 8.0, and 0.1% Tween 20.

METHODS

Passaging of Primary Human and Mouse Cells

1. Thaw a vial of early passage cells, by holding at 37°C in a water bath.
2. Prepare DMEM and pre warm it to 37°C.
3. Add 8 mL warm DMEM to cell pellet in a 15-mL tube.
4. Centrifuge at 800 rpm for 5 min to pellet cells.
5. Aspirate the media and discard.
6. Resuspend the pellet in 10 mL of the respective media. Transfer to a 10-cm tissue culture plate and incubate 37°C, 5% CO_2.
7. When the cells are approx 60% confluent (about 3 d),passage the cells 1 in 3.
8. Aspirate the media and discard (tilt plate to ensure complete removal).
9. Wash the plate with 2 × 5 mL PBS. Do this by pipetting onto the side of the plate, not the base, so as not to dislodge cells. Swirl gently, then aspirate; repeat.
10. Add 1 mL Trypsin/EDTA solution diluted 1:1 in PBS and swirl to cover. Incubate at 37°C for 10–20 min depending on the cell type.
11. Check plates under a microscope to ensure that cells are fully trypsinized. Cells should be rounded and float freely.
12. Centrifuge at 800 rpm for 5 min to pellet cells.
13. Aspirate the media and discard.
14. Resuspend the pellet in 10 mL of the respective media. Transfer to a 10-cm tissue culture plate and incubate 37°C, 5% CO_2.
15. Cells should be passaged every 3–5 d.

Determination of Replicative Life Span

In order to describe the growth characteristics of the respective cells prior to and after introduction of different genetic elements, one has to calculate the replicative life span of the individual cell types. Therefore, parental and genetically altered cells are routinely maintained in subconfluent conditions. The population doublings (PD) are determined using the following formula: PD = log(*N*/*N*0)/log2, where *N* is the number of collected cells and *N*0 is the number of seeded cells. Morphological changes are assessed with phase contrast microscopy, and scored by counting photographed five high-power fields (×200).

Retroviral Transduction

In order to introduce defined genetic elements capable of inducing a senescence phenotype in primary cells, a retrovirally mediated gene transfer is still the preferred method, although recent advances in lentiviral technology provide a compelling alternative. To produce replication deficient retroviruses, the amphotropic or ecotropic packaging cell line Phoenix A or Phoenix E, respectively are grown in supplemented DMEM and then transiently transfected with the respective retroviral vector to generate amphotropic or ecotropic replication-incompetent retroviruses, respectively.

Genes of interest can be expressed using, e.g., the pBabe vector system, the WZL-Hygro-based retroviral vectors, or the pFBneo retroviral vector system (Stratagene). Thereby oncogenic Ras (H-RasV12) was, for example, expressed using the pBabe-Puro vector. DN-hTert was created by substituting the aspartic acid and alanine residues at positions 710 and 711 with valine and isoleucine, respectively, by site-directed mutagenesis of the pCI-neo-hTERT-HA and subsequently subcloned in a retroviral vector. Alternatively, inducible vector systems can be used successfully.

Retroviral supernatant is harvested 48 h after transfection and filtered through a 0.45-μm filter, and fresh retroviral supernatant is used for infection of exponentially growing cells. Medium containing the appropriate antibiotic is exchanged to start selection 48 h after infection. In each set of experiments, multiple clones are pooled for further processing. For infection, 1 × 10^5 cells are seeded per well of a six-well tissue culture plate and incubated overnight.

Cells are exposed to 0.4 mL of the virus-containing medium added into 2 mL of fresh medium per well in the presence of 4 μg/mL of polybrene. Day 4 postinfection is generally designated as day 0. In all experiments, cells infected with an empty vector are used as control. After infection, successfully transduced polyclonal cell populations are obtained by selection with the appropriate drug (hygromycin [50 μg/mL], G418 [200 μg/mL], puromycin [0.5 μg/mL], blasticidin [2.5 μg/mL], or 500 μg/ml zeocin). Infection frequencies are typically 20–30%. Here, the optimal conditions for epithelial cell transduction are summarized:

1. Use amphotropic virus for human cells. Use ecotropic virus for mouse cells.
2. Virus titer or medium volume to be added: 20% of medium volume for primary cells grown in serum free medium ($\sim 1 \times 10^7$ infectious particles).
3. Infection methods: spin infection (1,800 rpm, room temperature, 45–90 min) (may be repeated) or normal infection.
4. Cell density upon infection: approx 20% confluency.
5. Polybrene concentration: 4–8 μg/mL.

Adenoviral Vectors and Infection

As an alternative to retroviral mediated gene transfer, and to ensure a high titer viral supernatant, adenoviral infection can be utilized. Recombinant adenoviral vectors expressing the gene of interest, e.g., p16INK4a, are generated. p16 cDNA is cloned into shuttle vector pAD/CMV and cotransfected with E3-deleted human adenovirus *dl*70001 into 293 cells, using calcium phosphate precipitation. Positive plaques are then identified by Southern blotting, repurified three times, and propagated in 293 cells. Subsequently, the recombinant virus is produced on a large scale and purified twice by cesium chloride density gradient centrifugation. Preparation of Ad5CMV/lacZ (Ad5lacZ) can be performed accordingly in order to measure infection efficiency. Target cells (3×10^5) are plated on 6-cm dishes 24 h before infection. Infection is then performed by adding 100 multiplicities of infections of adenovirus.

Lentiviral Production, and Infection

For production of lentivirus, 293T cells (human embryonic kidney cells) are transfected by the calcium-phosphate method using 10 μg transfer vector HIV-CS-CG or pLENTI-SUPER, 3.5 μg of VSVg envelope vector pMD.G, 2.5 μg of RSV-Rev, and 6.5 μg of packaging vector pCMVDR8.2. Lentiviruses are harvested 24 and 48 h after transfection and filtered through a 0.45-μ*M* filter. For example, wild-type MEFs are cultured to passage 9–10 whereupon cells are counted every 3–4 d 14 d prior to lentiviral infection. 1.8×10^5 senescent wild-type MEFs in 6 cm dishes were infected with lentivirus for at least 12 h in the presence of 8 μg/mL polybrene and were then allowed to recover for 48 h before reseeding for growth curves.

The following paragraphs include the description of procedures to characterize the growth of the respective cells, including cell cycle analysis, proliferation, senescence itself as well as telomere biology.

Flow Cytometry

Different primary and transduced cells are collected at different *population doubling* (PD) times and centrifuged at 300*g* for 4 min. The cell pellet is resuspended in 0.5 mL of PBS, fixed in 5 mL 70% ethanol, and stored at -20°C overnight. The cells are then washed twice with PBS and resuspended in a 1-mL solution containing 3.8 m*M* sodium citrate and 10 μg/mL propidium iodide. After 10 mg/mL of RNase treatment at 37°C for 20 min, the samples are analyzed by a fluorescence-activated cell sorter.

[^{3}H]Thymidine Incorporation Assay

Cells (0.25 × 10^5) are seeded per well in 12-well tissue culture dishes. Following retroviral infection, cells are incubated for 24 h with 1 Ci of [methyl-3H] thymidine (1 Ci/mL) added into each well. To harvest cells, cells are washed three times with cold phosphate-buffered saline, once with cold 10% trichloroacetic acid, and three times with 5% trichloroacetic acid. Cells are then lysed with 0.5 N NaOH on ice for 10 min, and neutralized with 0.5 *N* HCl. The cell lysate is supplemented with 10% trichloroacetic acid, incubated on ice for 20 min, and filtrated with glass microfiber filters using a sampling vacuum manifold. The filters are rinsed three times with ethanol, dried at 80°C for 45 min, and suspended in 10 mL of Ready Value Liquid Scintillation Cocktail to measure radioactivity with an LS 6500 Multi-Purpose Scintillation Counter.

SA-β-Gal Staining

To determine senescent cells, the assay for senescence associated β-galactosidase is carried out, e.g., with the commercially available Senescence associated β-Galactosidase Staining Kit. Cells are washed with PBS, fixed in 0.5% glutaraldehyde, and are incubated with freshly prepared SA-β-Gal staining solution for approx 12 h at 37°C in the dark. Five independent stainings are carried out for both early- and late- passage cells. The number of SA-β-Gal-positive cells in 10 low-power fields (×10) is then counted randomly and expressed as a percentage of all cells counted under phase contrast microscopy.

Western Blotting

Cells are washed with ice-cold phosphate-buffered saline and lysed with RIPA buffer. After 30 min on ice, the cell lysates are cleared by centrifugation at 14,000 rpm at 4°C for 15 min. Protein concentration is determined by the BCA protein assay. Protein (10–20 μg) is heat-denatured in NuPAGE lithium dodecyl sulfate sample buffer containing

NuPAGE reducing agent at 70°C for 10 min. The protein samples are separated on a 4–12% SDS-polyacrylamide gel electrophoresis (PAGE) and transferred to a polyvinylidene difluoride membrane (Immobilon-P; Millipore). The membrane is blocked in 5% nonfat milk (Bio-Rad) in TBST for 1 h at room temperature. Membranes are probed with the respective primary antibody diluted in 1% TBST milk overnight at 4°C, washed three times in TBST, incubated with an anti-mouse or anti-rabbit horseradish peroxidase antibody diluted 1:3000 in TBST for 1 h at room temperature and then washed three times in TBST. The signal is visualized by an enhanced chemiluminescence solution and is exposed to Eastman Kodak Co. X-Omat LS film. Densitometry of Western blots is performed using Scion Image Beta 4.02 Software. Data is calibrated with β-actin or β-tubulin as a loading control.

Telomerase Activity and Telomere Length Assays

Cellular extracts of all generated cell are assayed for telomerase activity using the PCR-based telomerase repeat amplification protocol (TRAP) assay. Hereby, telomerase within cellular extracts adds in a first step a number of telomeric repeats onto the 3' end of a substrate oligonucleotide. In a second step, the extended products are amplified by PCR generating a ladder of products with 6 base increments starting at 50 nucleotides. Cellular extracts (50 and 500 ng), along with heat-inactivated controls, are used for TRAP assays.

The measurement of telomere length is generally done by Southern blot hybridization, which requires unfragmented and pure, extracted genomic DNA. The most commonly used enzymes for the restriction of telomeric DNA are *Hinf*1 and *Rsa*1. Ten micrograms of the fragments obtained by digestion of genomic DNA with these restriction enzymes are resolved by electrophoresis, hybridized to a ^{32}P-labeled telomeric (CCCTAA) probe and the terminal restriction fragment (TRF) values are obtained by densitometric analysis. The resulting telomere restriction fragment band represents the mean telomere length of all chromosomes. In order to evaluate telomere length on individual chromosomes fluorescence *in situ* hybridization (FISH) techniques as quantitative FISH can be used.

Notes

1. For Kerationocyte culture, the calcium concentration is essential, because keratinocytes are sensitive to calcium induced differentiation. Therefore, a final Ca^{2+}-concentration of 0.4 m*M* should be maintained.

2. Defined Keratinocyte SFM, supplemented with defined Growth Factors and 1% L-glutamine and 1% penicillin/streptomycin gives better growing conditions for oral keratinocytes as compared to regular Keratinocyte SFM.
3. Calculate the exact concentration of BPE every time since concentrations between charges provided can vary.
4. Do not centrifuge primary cells above 800 rpm.
5. Primary keratinocytes should not be cultured to more than 30–40% confluence because they are sensitive to differentiation.
6. The use of PBS without calcium and magnesium is important in culturing primary keratinocytes because these cells are sensitive to calcium.
7. A 1:1 solution of trypsin/EDTA to PBS has the best efficiency and least toxicity.
8. If reducing the medium volume at time of transfection to, e.g., 5 mL/10 cm plate will increase the virus titer for unexplained reasons.
9. For some cells, spin-infection works better than incubation only.
10. For some cell types, a polybrene concentration of 8 μg/mL gives higher infection efficiency.
11. Key factors to consider for successful retroviral transduction: virus titer or medium volume to be added, infection methods, simple incubation vs spin infection, cell density on infection, polybrene concentration.
12. Drug selection doses should be determined empirically for each drug and each cell type used. The doses mentioned are approximate doses proved to work in different primary culture conditions, which can be considered as starting points.
13. β-Gal staining solution should be freshly prepared before use, because this increases staining dramatically. Additionally, X-Gal is not stable in water and should be added to the staining solution just before use. Five-millimolar potassium ferrocyanide, 5 m*M* potassium ferricyanide are biohazards. Consult the MSDS for proper handling instructions.
14. TEMED is best stored at room temperature in a desiccator. Buy small bottles, as it may decline in quality after opening.

Methods of Cellular Senescence Induction Using Oxidative Stress

Cellular senescence was first described by Hayflick and colleagues in 1961 when they observed that human diploid fibroblasts underwent a finite number of cell divisions in culture. Subsequently, it was found that most types of normal cells in primary culture will ultimately exhaust their replicative capacity and enter into a growth arrest state termed replicative senescence. This is associated with telomere shortening, which has been attributed to the DNA end-replication problem in the absence of telomerase. Senescent cells become irreversibly growth arrested yet remain viable for extended periods of time. These cells undergo a dramatic change in morphology, becoming enlarged and flattened in shape, and also express the senescence-associated β-galactosidase (SA-β-gal). At the molecular level, senescence is associated with changes in the expression of a large number of genes. Noticeably, two pathways that are involved in cell cycle regulation, the p53 pathway and the Rb pathway, play important roles in replicative senescence.

It is now clear that replicative senescence is just one example of a more widespread response termed cellular senescence. Studies have shown that the senescent phenotype can be induced prematurely in early passage cells by agents that cause DNA damage or disrupt heterochromatin, or by strong mitogenic signals. One potent process capable of producing DNA damage is oxidative stress. These forms of premature senescence are often induced within a period as short as several days and are not accompanied by significant telomere shortening. Nevertheless, premature senescence can also be achieved in a telomere-dependent manner by exposure to mild oxidative stress. This is because mild oxidative stress causes single strand breaks in telomeric regions, which consequently cause accelerated telomere shortening.

Cellular senescence induced by oxidative stress has been regarded as an excellent in vitro model for aging research. The free radical theory of aging predicts that *reactive oxygen species* (ROS) produced during normal aerobic metabolism cause damage to biomolecules which ultimately results in decline of tissue functions and aging. ROS are oxygen-centered free radicals, such as superoxide and hydroxyl radicals as well as other reactive nonradical species such as hydrogen peroxide and singlet oxygen. Much evidence is available to support this theory, and thus the role of oxidative stress in aging has been extensively studies. Various oxidative stresses have been used to induce premature senescence, including exposure to hydrogen peroxide, ultraviolet (UV) light, *tert*-butylhydroperoxide, and hyperoxia, among which hydrogen peroxide is the most commonly used inducer. Here, we describe detailed protocols for induction and detection of cellular senescence using hydrogen peroxide.

Materials

Cell culture

1. 18 × 18 mm, thickness number 1 glass coverslips (BDH): wash the coverslips once in distilled deionized water, once in absolute ethanol, air-dry on a sheet of absorbent paper, place them in a clean glass beaker, cover with foil and autoclave tape, and autoclave. Open only under sterile conditions.
2. Cryogenic vials (NALGENE).
3. Human fibroblasts strain IMR-90.
4. Modified Eagle's minimal essential medium (EMEM; ATCC) supplemented with 10% fetal bovine serum (FBS; ATCC), 100 U/mL penicillin, and 0.1 mg/mL streptomycin (penicillin-streptomycin solution).
5. Dulbecco's phosphate-buffered saline (PBS) (without Ca^{2+} and Mg^{2+}, Sigma).
6. 1X Trypsin-EDTA in PBS (10X Trypsin-EDTA solution from Invitrogen).
7. Dimethyl sulfoxide (DMSO; Sigma).
8. Hydrogen peroxide (30% [W/W] solution from Sigma).

SA-β-gal assay

1. Light microscope.
2. PBS: prepared using Sigma PBS tablets.
3. 0.1 *M* citric acid solution: 2.1 g citric acid monohydrate in 100 mL distilled deionized water.
4. 0.2 *M* sodium phosphate solution: 2.84 g/100 mL sodium phosphate dibasic or 3.56 g/100 mL sodium phosphate dibasic dihydrate.
5. Citric acid/sodium phosphate buffer, pH 5.0 or 6.0: by mixing 0.1 *M* citric acid solution to 50 mL of 0.2 *M* sodium phosphate solution until the pH reaches the desired value.
6. 200 mg/mL X-gal solution: dissolve 100 mg X-gal powder in 0.5 mL dimethylformamide (DMF); stored in dark at –20°C.
7. 5 *M*NaCl.
8. 100 m*M* potassium ferrocyanide solution: 2.12 g in 50 mL distilled deionized water.
9. 100 m*M* potassium ferricyanide solution: 1.65 g in 50 mL distilled deionized water.

10. 1 *M* magnesium chloride solution: 10.17 g in 50 mL distilled deionized water.
11. SA-β-gal staining solution (to be freshly made): for 20 mL staining solution, combine 4 mL of citric acid/sodium phosphate buffer with 100 μL X-gal solution. Then add 1 mL of 100 m*M* potassium ferrocyanide, 1 mL of 100 m*M* potassium ferricyanide, 0.6 mL of 5 *M* NaCl, 40 μL of 1 *M* magnesium chloride, 13.26 mL distilled deionized water.
12. 4% paraformaldehyde fixative solution: dissolve 4 g of paraformaldehyde in 50 mL of distilled deionized water, and then add 1 mL of 1 *M* NaOH solution, stir the mixture gently on a heating block (~65°C) until the paraformaldehyde is dissolved. Next, add 10 mL of 10X PBS and allow the mixture to cool to room temperature. Adjust the pH to 7.4 using 1 *M* HCl (~1 mL is needed) and then adjust the final volume to 100 mL with distilled deionized water. Filter the solution through a 0.45-μm membrane filter to remove any particulate matter, and store in aliquots at –20°C.

BrdU incorporation assay

1. Fluorescence microscope with filters suitable for detection of fluorescein isothiocyanate (FITC) and TOTO-3-iodide.
2. Labeling reagent: 5-bromo-2' -deoxyuridine (BrdU, SIGMA), 5 mg/mL (16.3 m*M*).
3. Permeablization solution: 0.2% (w/v) Triton X-100 (SIGMA) in PBS.
4. Denaturing solution: 2 *M* HCl.
5. Mouse monoclonal antibody to BrdU conjugated with fluorescein.
6. TOTO-3-iodide.

Immunofluorescence microscopy

1. Confocal laser microscope.
2. 10X PBS.
3. 4% paraformaldehyde.
4. Tris-buffered saline (TBS): 50 m*M* Tris-HCl (pH 7.4), 150 m*M* NaCl.
5. Quench solution: 0.1% sodium borohydride in PBS (dissolve in PBS on day of use, wear appropriate gloves and masks and handle with great care).
6. 0.2% Triton X-100 in PBS.

7. Bovine serum albumin (BSA).
8. Blocking buffer: 10% horse or goat serum, 1% BSA, 0.02% NaN_3, 1X PBS.
9. Antibody dilution buffer: 1% BSA in TBS.
10. Antibodies against γH2AX (Upstate) and 53BP1.
11. Secondary antibodies: Alexa Fluor 488 donkey anti-mouse immunoglobulin (Ig)G and Alexa Fluor 555 goat anti-rabbit IgG.
12. Mounting medium: Antifade.

Methods

When fibroblast cells are newly purchased from a commercial source, they should be amplified (≤ 5 population doublings [PDs]) and cryopreserved. Protocols described in this chapter have been optimized for human fibroblast IMR-90 cells. If different cell strains are to be used, dose–response experiments should be carried out to obtain the most effective H_2O_2 concentration without killing the cells. As a guide titrations in the range of 50 μM to 800 μM should be carried out.

As a small fraction of cells can recover from a single H_2O_2 stress and re-enter cell cycle, they will multiply and eventually overtake the induced premature senescent cells until contact inhibition occurs. This can be largely avoided by treating the cells with a second H_2O_2 stress. Morphological changes of oxidatively stressed cells take place gradually when plated at a low density to enable the cells to fully enlarge. Eventually, they will display morphological features that are indistinguishable from replicatively senescent cells. In addition to the morphological changes, senescent cells are also detected by permanent cell cycle arrest and the appearance of SA-β-gal activity. No incorporation of BrdU into nuclear DNA after 48 h pulse labeling is usually regarded as a sufficient criterion for permanent growth arrest in a senescence study, although longer BrdU labeling could be adopted. SA-β-gal staining at pH 5.0 and pH 6.0 for early passage IMR-90 cells and SA-β-gal staining at pH 6.0 for H_2O_2-induced premature senescent cells to show the outcome of the assay. H_2O_2 treatment causes persistent DNA double-strand breaks which can be detected by immunofluorescent staining using antibodies recognizing DNA double-strand break marker proteins, e.g., γH2AX and 53BP1. Persistent DNA damage caused by oxidative stress may be responsible for the induction of premature senescence.

The effect of oxidative stress can also be assessed by measuring increases in levels of p53, p21, and γH2AX as well as hypophosphorylation of retinoblastoma protein (Rb). This can all be determined by Western blotting using appropriate antibodies.

Freezing cells

1. Check the culture for (a) healthy growth and (b) no contamination.
2. Grow the culture up to the late log phase and trypsinize and count the cells.
3. Pool the trypsinized cells in a 50-mL Falcon tube and centrifuge for 2 min at 160*g*.
4. Dilute DMSO in growth medium to 5% to make freezing medium.
5. Discard supernatant and resuspend the cells in freezing medium at approximately 1×10^6 cells/mL.
6. Dispense 0.5-mL aliquots into prelabeled cryogenic vials.
7. Slow-freeze the cells in the vials between sheets of paper towels in a polystryrene box at –80°C in a regular deep freezer overnight.
8. Transfer the vials to liquid N_2 the following morning.

Thawing frozen cells

1. Prewarm 17 mL of growth medium in a prelabeld 75-cm^2 culture flask in a 37°C incubator.
2. Retrieve the vial from the liquid N_2 and thaw it quickly between the hands.
3. When thawed, spray the vial with 70% ethanol and transfer the contents of the vial to the prewarmed culture flask.
4. Change the medium the following morning to remove residual DMSO.

H_2O_2 treatment

1. Trypsinize the exponentially growing cells, count and re-seed 1.8×10^6 cells in 12 mL of growth medium/dish in 100-mm dishes.
2. Leave the cells to settle for 24 h.
3. Prepare an appropriate amount of 600 μM H_2O_2 in growth medium.
4. Take the dishes out of the incubator and remove the medium.
5. Add 13 mL of 600 μM H_2O_2 per dish.
6. Return the cells to the incubator.
7. Remove the medium after 2 h, re-feed with medium containing no H_2O_2, and incubate for 4 d.
8. Split the cells in a 1:2 ratio and incubate for 24 h.
9. Treat the cells with H_2O_2 for a second time by repeating steps 3-6.

10. Check daily to monitor morphological changes of the cells. The shows morphologies of early passage control and replicatively senescent human fibroblasts.

BrdU incorporation assay

1. Transfer cells grown on coverslips into a six-well plate containing 2 mL of medium/well.
2. Add BrdU to 10-μM final concentration.
3. Label for 48 h.
4. Aspirate the medium. Wash the cells with two changes of PBS.
5. Aspirate the last change of PBS. Add 2 mL of 4% paraformaldehyde. Fix at room temperature for 5 min.
6. Wash three times in PBS.
7. Permeablize the cells in 2 mL of 0.2% Triton X-100 in PBS for 5 min.
8. Wash two times in PBS.
9. Aspirate the last wash. Add 2 mL of 2 *M* HCl and incubate at room temperature for 1 h to denature the DNA.
10. Wash three times in PBS. Leave the third wash in contact with coverslips while preparing 1:20 diluted anti-BrdU containing 0.5 μM TOTO-3-iodide in PBS.
11. Place coverslips between two wells of a 12-well plate. Add 150 μL of anti-BrdU solution containing TOTO-3-iodide onto the coverslips. Incubate in humidified dark chamber at room temperature for 1 h.
12. Wash three times in PBS and once in distilled water (this removes salts from the coverslips).
13. Analyze the coverslips with the fluorescence microscope. Count the total number of nuclei that were stained with TOTO-3-iodide within the field of view.
14. View the same field under FITC filter. Count the positive cells within the field of view.
15. Repeat steps 13 and 14 until the count has reached 500 total nuclei.

SA-β-gal assay

1. Wash the cells twice in PBS.
2. Fix with 4% paraformaldehyde for 5 min at room temperature. Prolonged fixation will inactivate the enzyme.
3. Wash three times in PBS.
4. Add freshly prepared staining solution (2 mL/well of six-well plate).

5. Incubate at 37°C in a humidified chamber overnight. Do not use an incubator with a 5–10% CO_2 environment such as a CO_2 incubator used for cell culture.
6. Examine the blue staining using a light microscope.

Immunofluorescence analyses for DNA damage

1. Treat cells as desired.
2. Wash cells on coverslips once with TBS.
3. Fix with 4% paraformaldehyde for 10 min at room temperature.
4. Wash coverslips once with TBS. Aspirate completely and permeabilize cells on coverslips with 0.2% Triton X-100 for 5 min at room temperature.
5. Wash coverslips three times for 5 min each with TBS at room temperature.
6. Quench cells in fresh 0.1% sodium borohydride in TBS for 5 min. Aspirate off completely and wash once in TBS.
7. Incubate coverslips with blocking buffer at room temperature for 1 h.
8. Wash once in TBS for 5 min.
9. Dilute the primary antibodies as appropriate in 1% BSA in TBS.
10. Place the coverslips between two wells of a 12-well plate and apply 220 μL of the antibodies/coverslip to the cells on coverslips.
11. Incubate at 4°C in a humidified chamber overnight.
12. Wash all coverslips three times for 5 min each with TBS.
13. Incubate coverslips with appropriate dilution of the fluorescence-labeled secondary antibodies in 1% BSA in TBS for 45 min at room temperature in the dark.
14. Wash coverslips three times for 5 min each with TBS. in low lighting.
15. Air-dry the coverslips in the dark.
16. Mount coverslips on slides using the ProLong Antifade Kit. Air-dry in the dark and seal the sample with nail varnish. The sample can be analyzed immediately by confocal microscope or be stored in the dark at 4°C up to a month.

Notes

1. Standard culture conditions with ambient oxygen concentrations of approx 20% are used. Compared to physiological oxygen levels of approx 3%, these culture conditions therefore expose cells to oxidative stress. This may be avoided by culturing cells under 3%

oxygen in a triple gas incubator. It is well documented that when cultured in 3% oxygen human fibroblasts achieve 20–30% more population doublings than those cultured under standard 20% oxygen.

2. A single-cell suspension is desirable to ensure an accurate count and uniform growth on reseeding. This can be achieved by pipetting the cell suspension up and down a few times after adding the medium to the trypsinized, detached cells, taking care not to create foam. Dilute the cell suspension to 1.5×10^5 cells/ml and seed 12 mL in a 100-mm dish to achieve 1.8×10^6 cells/dish. If the BrdU assay or SA-β-gal assay are to be conducted for the cells, coverslips may be placed in the dishes prior to cell-seeding.
3. Always prepare the reagent freshly. Take the following formula as a guide to calculate the amount of 30% H_2O_2 solution in μL to make 600 μM H_2O_2 in X mL of medium: $(600 \times X \times 34)/0.3/10^6$. It should be noted that the biological activity of H_2O_2 is largely dependent on the medium, i.e., serum-containing medium < serum-free medium < buffer. Medium containing 10% FBS is used for H_2O_2 treatment in this protocol in order to eliminate any effects other than H_2O_2.
4. It is important to apply the medium containing H_2O_2 gently and evenly onto the cells by dripping the medium with a circular movement from the centre to the side of the dish. Adding all 13 mL onto one spot may cause local killing of the cells.
5. As reported, it is difficulty to achieve a 100% cell cycle arrest and induction of senescence by a single H_2O_2 treatment, because a fraction of the treated cells can recover from the oxidative stress and re-enter cell cycle. Therefore, it is necessary to split the cells and treat with H_2O_2 for a second time. Again, cover-slips may be placed in one dish for BrdU and SA-β-gal assays that can be carried out after the second H_2O_2 treatment.
6. The treated cells will gradually get enlarged but the characteristic fully enlarged morphology can only be seen when cell density is low enough for the cells to stretch.
7. If the analysis can be carried out immediately after staining, it is usually unnecessary to mount the coverslip. Otherwise, the coverslip should be mounted using appropriate mounting medium and stored at 4°C. We use the ProLong Antifade Kit from Molecular Probes.
8. It is important that nonconfluent cultures be used because nonsenescent cells that are confluent or maintained at confluence may show false-positive staining.

9. It is useful to include a positive control for the assay. This is achieved by shifting the pH of the staining solution to pH 5.0 by using citric acid/sodium phosphate buffer at pH 5.0. This detects all mammalian cells with endogenous β-galactosidases.
10. Centrifuging the antibodies for 20 min at 12,000*g* in a refrigerated microcentrifuge prior to use will reduce any aggregated material, thereby reducing background. When using any primary or fluorescence-labeled secondary antibody for the first time, titrate out the antibody to determine which dilution allows for the strongest specific signal with the least background for the sample.
11. DAPI (final concentration 0.1 μg/mL) can be included in this step. Alternatively, DAPI can be added to the mounting medium.
12. As directed by Molecular Probes (#P-7481), prepare just before use. Add 1 mL of ProLong mounting medium to one vial of ProLong antifade reagent. Mix gently. Any unused mixture can be stored at –20°C for up to 1 mo.

3

LIFE SPAN

Normal human somatic cells undergo a finite number of replications and enter irreversible arrest of cell divisions through a regulated senescence process. Cellular senescence is a physiologically important event for both aging and cancer. Senescent cells have been reported to accumulate *in situ* during aging and in tissues with age-associated pathologies. Cellular senescence is also a potent tumor suppressive mechanism which must be overcome by aberrant cells for cellular immortalization and tumorigenic conversion. Although the detailed molecular mechanisms of senescence remain to be determined, telomere length and status are known to limit the replication of normal cells. Telomeres are hexameric repeat sequences flanking the chromosomal ends and are essential for maintaining the stability of the cellular genome. Telomeres progressively shorten during DNA replication as a result of an end-replication problem in the absence of telomerase, an enzyme which catalyzes *de novo* synthesis of telomeric DNA.

Telomerase is a ribonucleoprotein complex composed minimally of *hTERT* (catalytic protein subunit) and *hTR* (RNA template). Telomerase enzyme activity is closely correlated with the expression level of *hTERT* and is reconstituted in normal human cells by ectopic expression of *hTERT* alone. Telomerase reconstitution has been demonstrated in a number of different normal human cell types, such as *human diploid fibroblasts* (HDF), human myoblasts, preadipocytes, lymphocytes, and keratinocytes. Ectopic *hTERT* expression in most normal human somatic cells resulted in significant extension of normal replicative life span, but the ability of *hTERT* to lead to cellular immortalization appears to be cell type-specific. For instance, immortalization of HDF is sufficed by ectopic expression of *hTERT* alone, whereas that of human keratinocytes required additional inactivation of the p16^{INK4A}/pRb tumor suppressor pathway. Immortalization of HDF with *hTERT* overexpression also occurs with crisis stage during which many cells undergo cell death or permanent replication arrest. Also, prolonged culture of *hTERT*-immortalized HDF is associated with alteration of the p16^{INK4A} and p53 tumor suppressor genes and exhibition of transformed phenotype. These results indicate that telomere shortening and altered status are not the only limiting factors of human somatic cell replication although telomerase activity is required for immortalization.

This article describes detailed methods to extend the replicative life span of normal human cells by telomerase reconstitution in cells as exemplified in HDF and *normal human keratinocytes* (NHK). In particular, we will discuss the methods to establish primary HDF and NHK cultures from excised tissues, reconstitution of telomerase activity by retrovirus-mediated *hTERT* expression, documentation of cellular replication kinetics, and assay of telomerase activity in vitro and in vivo.

Materials

Cell Culture Media and Reagents

1. Dulbecco's modified Eagle's medium (DMEM) supplemented with 10% fetal bovine serum (FBS). All cell culture media is stored at 4°C.
2. Keratinocyte basal medium (KBM) supplemented with growth factor bullet kit. The KBM plus the growth factor kit is called *keratinocyte growth medium* (KGM), which should be maintained at 4°C. Before reconstitution, the growth factor kit is stored at -20°C.
3. Cell freezing medium: normal nutrient medium containing 50% FBS (HyClone) and 10% dimethylsulfoxide (DMSO).
4. GP2-293 universal packaging cell line.
5. pVSV-G envelope plasmid.
6. Calcium-phosphate transfection kit (Invitrogen) containing sterile water, 2*M* $CaCl_2$, and 2X HEPES-buffered saline (HBS).
7. S-minimal essential medium (S-MEM) (Invitrogen) supplemented with 10% FBS (HyClone).
8. Solution of trypsin (0.25%) and ethylenediamine tetraacetic acid (EDTA) (1 m*M*) from Invitrogen.

9. 1X phosphate-buffered saline (PBS): 8 g NaCl, 0.2 g KCl, 1.44 g Na_2HPO_4, 0.24 g KH_2PO_4 in 1 L tissue culture-grade water. The pH is adjusted to 7.4 with HCl. This solution is conveniently made in 10X concentration, which can be diluted 10-fold before use.
10. Geneticin G418 stock solution (200 mg/mL) from Invitrogen.
11. Puromycin (Sigma) dissolved in tissue culture grade water at 1 mg/mL.
12. Hygromycin (Sigma) dissolved in tissue culture grade water at 10 mg/mL.
13. Collagenase (type II 381 U/mg solid; Sigma).
14. Dispase II (neutral protease from Bacillus polymyxa). Both collagenase and dispase II are needed for establishing the primary cultures of NHK. The collagenase and dispase II solutions need to be made fresh before use.

Telomerase Enzyme Assay

1. TRAPeze XL Telomerase Detection Kit.
2. *Taq* DNA polymerase.
3. T4 polynucleotide kinase (PNK).
4. Adenosine 5'-triphosphate [α-^{32}P] (4500 Ci/mmol; 10 mCi/mmol). The radioactive materials are hazardous and require special handling and disposal according to the guidelines set forth by the institutional Radiation Safety Office.
5. 1X TRAP reaction buffer (20 m*M* Tris-HCl, pH 8.3, 1.5 m*M* $MgCl_2$, 63 m*M* KCl, 0.005% Tween 20, 1 m*M* EGTA).
6. 1X CHAPS buffer (10 m*M* Tris-HCl, pH 7.5, 1 m*M* $MgCl_2$, 1 m*M* EGTA, 0.5% 3-[(3-cholamidopropyl)-dimethylammonium]- 1-propanesulfonate, 10% glycerol, 5 m*M* β-mercaptoethanol, 0.1 m*M* benzamidine).
7. 5X Protein Assay solution.
8. Twelve percent polyacrylamide gel made from 30% acrylamide/ bis solution, 10% ammonium persulfate (APS), and *N,N,N,N'* -Tetramethyl-ethylenediamine.

Terminal Restriction Fragment Length Assay

1. Cell lysis buffer: 10 m*M* Tris-HCl (pH 8.0), 100 m*M* EDTA (pH 8.0), 0.5% sodium dodecyl sulfate (SDS).
2. 20X SSPE transfer buffer: 175.3 g of NaCl, 27.6 g NaH_2PO_4-H_2O, and 7.4 g EDTA in 1 L total volume with H_2O.

3. Proteinase (endopeptidase) K dissolved in water at 100–500 U/mL concentration.
4. DNase-free ribonuclease (RNase) A from bovine pancreas (Sigma) dissolved in water at 10 mg/mL concentration. To ensure absence or inactivation of DNase in this enzyme solution, a vial of RNase can be boiled for 10 min because DNase is heat-labile but RNase is heat-resistant.
5. Ultra-Pure buffer-saturated phenol.
6. *HinF*I and *Rsa*I restriction endonucleases.
7. Hybond nylon nucleic acid transfer membrane.
8. $(TTAGGG)_4$ synthetic oligonucleotide.
9. T4 polynucleotide kinase (PNK).
10. Adenosine 5'-triphosphate [α-^{32}P] (4500 Ci/mmol; 10 mCi/mmol).
11. Hybridization buffer: 0.5 *M* Na_2HPO_4 pH 7.2, 7% SDS, 1% bovine serum albumin (BSA), 0.5 m*M* EDTA.

Methods

Telomerase reconstitution by ectopic expression of *hTERT* can be achieved either by transfection of cells with mammalian expression vector or infection with retroviral vector carrying full-length *hTERT* cDNA. The retrovirus-mediated gene transfer in general allows for more predictable and homogenous expression of the transgene than does transfection, particularly in primary human epithelial cells. Using plasmid transfection method, it is also possible to lose the transgene expression after long-term culture of cells. Therefore, for the purpose of long-term expression of the transgene, retroviral vectors are recommended. In this section, we discuss the methods of establishing the primary human cells (HDF and NHK), preparation of pantropic retroviral vectors capable of expressing *hTERT*, infection of cells with the viral vector, and detection of reconstituted telomerase activity in vitro and in vivo.

Establishing Primary Cultures of HDF and NHK

Primary HDF culture establishment from tissue explants

1. Monolayer HDF primary cells can be obtained by explantation of dermal or oral connective tissues immediately following excision of the tissues from the donors. The tissue specimens are washed thoroughly in DMEM without serum to remove any contaminants and minced to approx 2 mm in diameter with fine scissors.
2. The minced tissues are placed in 60-mm culture dishes with 2 mL DMEM containing 10% FBS and maintained in humidified

incubator with 5% CO_2 at 37°C. The nutrient medium is changed every 2 d. After 1–2 wk, replicating HDF cells are visible surrounding each tissue explant.

3. When the culture dish becomes 60-70% confluent, the tissue explants can be removed with sterile forceps and the cells transferred to new dishes with desired cell density by trypsinization -2.5 × 10^3 cells/cm^2 (2 × 10^5 cells per 100-mm dish) yields well dispersed culture which becomes 60–70% confluent within one week.
4. The primary HDF cells are then maintained in serial subcultures, documenting the precise number of cumulative population doublings (PDs).

Primary NHK culture establishment from epithelial tissues of skin or oral mucosa

1. The epithelial sheet can be separated from the underlying connective tissue by incubating the excised tissue in 10 mL S-MEM solution containing 25 mg dispase II and 3 mg collagenase for 90 min at 37°C.
2. After the enzyme treatment, the epithelial sheet can be easily removed from connective tissue using a pair of fine forceps.
3. The isolated epithelial sheet is minced finely with microscissors—the degree of mincing determines the amount of epithelial cells that can be harvested by trypsinization.
4. Single-cell suspension is obtained by treating the minced epithelial tissues with 0.25% trypsin for 5 min followed by neutralization with S-MEM containing 10% FBS. The duration of the trypsin digestion is limited to 5 min to avoid harvesting cells from the suprabasal layers. Also, trypsin digestion longer than 5 min greatly reduces the cell viability.
5. The isolated epithelial cells need to be washed once with KGM and plated in T25 tissue culture flask precoated with collagen. NHK cells are serially maintained in KGM, being passaged at every 60–70% confluency levels. In general, primary NHK cells replicate with a doubling time of approx 30 h and can be passaged four to five times before senescence. The number of cells plated can be varied depending on the purpose of each culture, but the cell density of 2.5 × 10^3 cells/cm^2 yields 60–70% confluency in 4–5 d for the exponentially replicating NHK cultures. For the purpose of retrovirus infection, it is very important to use actively replicating (primary or secondary) cultures of HDF and NHK.

Ectopic Expression of hTERT by Retrovirus-mediated Gene Transfer

Plasmid containing *hTERT* cDNA (pCI-hTERT) can be obtained from the laboratory of Dr. Robert A. Weinberg. To create a retroviral vector expressing *hTERT*, the *hTERT* cDNA must be subcloned into a retrovirus expression plasmid, e.g., pLXSN, pLPCX, pLHCX, or pLXRN, which is readily available from a commercial source. Retroviral vector for *hTERT* is then prepared in GP2-293 universal packaging cell line, which allows for pseudotyping with helper envelope plasmid. GP2-293 cells are transfected with a mixture of the retrovirus expression plasmid (containing the *hTERT* cDNA) and pVSV-G plasmid. Virus production from the packaging cells is allowed for up to 48 h, after which the virus is collected and concentrated by ultracentrifugation.

Preparation of retroviral vectors

1. One day before transfection, 1–3 × 10^6 GP2-293 packaging cells are plated per one 100-mm tissue culture dish—this should yield approx 70% confluency level on the following day.
2. Total of 20 μg (1 μg/μL) plasmid DNA (15 μg retrovirus expression plasmid and 5 μg pVSV-G) is added to 244 μL tissue culture grade water in 50-mL conical centrifuge tube. Then, 36 μL of 2*M* $CaCl_2$ is added to the DNA solution and incubated in ice for 5 min.
3. The DNA-$CaCl_2$ solution is mixed drop-by-drop with 300 μL of 2X HBS while mixing the solution after each drop of HBS. The solution containing the DNA-calcium-phosphate complex is lett at room temperature for 20 min and added onto the GP2-293 cells in a 100-mm dish with 10 mL of nutrient medium.
4. Transfection is allowed for 6 h, after which the culture medium is changed with fresh warm medium.
5. After 48 h, the cell-free culture supernatant containing the virus is centrifuged at 50,000*g* for 90 min at 4°C. The virus pellet is resuspended in 500 μL of desired culture medium and used immediately for infection or snap-frozen in liquid N_2 for long-term storage at –80° C.

Infection and selection of the cells

1. On the day before infection, the target cells (~3 × 10^5) are plated in a 60-mm culture dish to achieve 50% confluency on the following day.
2. The concentrated virus stock is thawed quickly in water bath at 37°C and placed in ice as soon as it is thawed.

3. After removing the medium from the target cell culture, 500 μL concentrated virus solution is placed onto the cells with 500 μL additional culture medium containing polybrene at 8 jig/mL final concentration.
4. Infection is allowed for 2 h, after which time the culture medium is changed with 4 mL fresh warm medium.
5. Forty-eight hrs after infection, the infectants can be selected with the desired antibiotics, e.g., puromycin, G418, or hygromycin, depending on the drug selection marker present in the retrovirus expression vector.
6. After selection is complete, the remaining drug-resistant cells are maintained in serial subcultures, documenting the precise number of cells plated and harvested per each passage.
7. The cells can be harvested for long-term storage in liquid N_2. Trypsinized cells ($\sim 10^6$) are resuspended in 1 ml cell freezing medium in a cryogenic vial and placed overnight at -80°C in cryo 1°C freezing container, which allows for gradual decrease in temperature. On the following day, the frozen cell vials are transferred to liquid N_2 for long-term storage. If necessary, the frozen cells should be thawed quickly in 37°C, washed once in normal nutrient medium, and plated in a tissue culture dish.

Determination of Telomerase Activity by Telomeric Repeat Amplification Protocol Assay

After infection of cells with the *hTERT* retroviral vector and selection with the drug marker, the presence of telomerase activity should be confirmed. Telomerase activity is determined by telomeric repeat amplification protocol (TRAP) assay, which is a PCR-based method that allows for rapid determination of telomerase activity in cells and tissues with remarkable sensitivity. TRAP assay consists mainly of two steps: telomere synthesis by telomerase and telomeric sequence amplification by PCR. TRAP assay can be performed using telomerase template, to which telomerase binds and adds tandem TTAGGG repeats, and the TS and CX primer set that can amplify the synthesized telomeric sequences by PCR. Alternatively, TRAP assay can be performed with commercially available TRAPeze XL Telomerase Detection Kit. TRAP assay is performed with the cultured cells according to the following protocol:

1. Cells are harvested by trypsinization and washed once with 1X PBS.

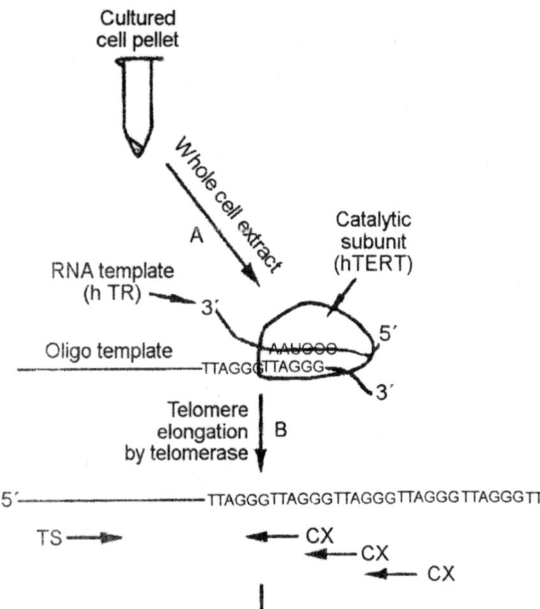

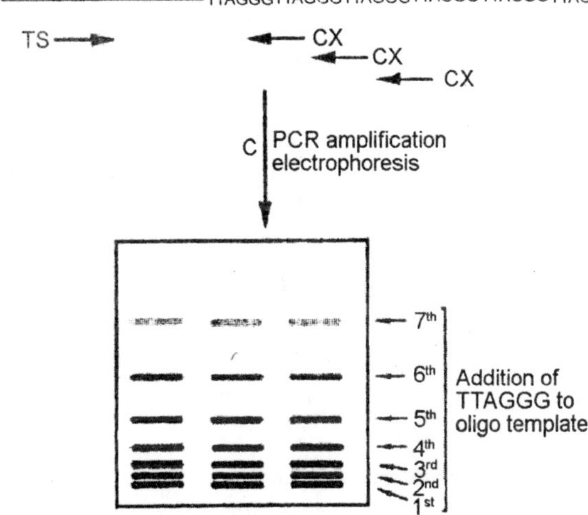

Fig. 3.1. TRAP assay is a PCR-based assay of telomerase activity. Cellular telomerase activity is detected by TRAP assay, which consists mainly of two steps: de novo synthesis of telomeres by telomerase and telomeric sequence amplification by PCR.

2. Washed cell pellet (from approximately half million cells) is then dissolved in 100 μL of 1X CHAPS buffer, incubated on ice for 30 min, and centrifuged at 12,000 rpm in a microcentrifuge at 4°C for 30 min.
3. The supernatant is aliquoted and frozen in liquid nitrogen for TRAP assay and for the determination of protein concentration.

4. The TRAP reaction is performed with the positive control cell extract, e.g., 293 cells, containing high telomerase activity and the negative control cell extract treated with 200 μg/mL RNase A. Alternatively, the 293 cell extract can be treated with heat (85°C for 10 min) to abolish the enzyme activity. Because telomerase enzyme activity depends on the integrity of the RNA template and is heat-sensitive, treatment of the cell lysate with either RNase or heat would suffice as negative controls.
5. Telomerase reaction mixture is prepared by adding 2 μL cell lysate containing 0.5 or 2.0 μg of cellular protein to 48 μL solution comprising of 1X TRAP reaction buffer, 50 *μM* each dNTP, 0.05 μg TS primer end-labeled with α-^{32}P-ATP (4500 Ci/mmol), 1 μL primer mix, and 0.4 U *Taq* DNA polymerase.
6. The mixture is incubated at 30°C for 30 min, and the telomerase reaction products are amplified in a DNA Thermal Cycler. The following conditions are used for the PCR cycles: 30 cycles at 94°C for 30 s and 55°C for 30 s, followed by one delayed extension cycle at 72°C for 10 min.
7. The PCR products are then resolved in 12% nondenaturing polyacrylamide gel in 1X tris-borate EDTA (TBE) buffer for 120 min at 30 W. After drying the gel, the radioactive signal can be detected by autoradiography using X-ray films or PhosphorImager.

Telomere Length Analysis by Southern Hybridization

After ectopic expression of *hTERT*, the telomere length becomes markedly elongated in cells to reflect in vivo function of reconstituted telomerase. The changes in the telomere length in cells can be determined by Southern hybridization described below.

Isolation of genomic DNA from cultured cells

1. Two to five million cells are lysed in 500 μL cell lysis buffer supplemented with 10 μL RNase A (10 mg/mL stock concentration—this is 100X concentrate) and proteinase K (final concentration of 100 μg/mL—this is usually 30 μL of the stock solution at 20 mg/mL). The lysis of cells and nuclear proteins can proceed at 50°C for at least three hours to overnight.
2. Phenol extraction is then performed with 500 μL buffer-saturated phenol. One needs to perform the phenol extraction twice with $CHCl_3$ and precipitate the isolated DNA in equal volume of isopropanol and one tenth volume of 3 *M* sodium acetate, pH 5.2. Isolated genomic DNA can be redissolved in 10 m*M* tris pH 8.0

after a single wash in 75% ethanol. Typical yield from two to five million cells is 30–50 μg genomic DNA.

Southern blotting and hybridization with telomeric probe

1. Ten micrograms of isolated genomic DNA is digested to completion with the restriction enzymes *HinF*I and *Rsa*I overnight.
2. DNA fragments are then electrophoresed in a 0.8% agarose gel and transferred to Hybond nylon membrane by capillary transfer method in 20X SSPE transfer buffer.
3. After the completion of transfer (~12–16 h), the membrane is washed briefly in 6X SSPE to remove any agarose particles and exposed to ultraviolet (UV) in a UV cross linker apparatus to immobilize the nucleic acids on the membrane.
4. Prehybridization is required to equilibrate the membrane and to reduce the non-specific background signals. It is usually performed in 20 mL solution of the hybridization buffer at 65°C for 4 h.
5. Hybridization is then performed with the telomere probe $(TTAGGG)_4$ end-labeled with α-^{32}P-ATP by PNK. Hybridization is usually allowed for overnight (12–16 h) at 65°C.
6. The filter is washed with 2X SSPE-0.1% SDS at room temperature for 15 min, and with 0.1X SSPE-0.1% SDS twice at 65°C for 15 min each.
7. The radioactive signals are scanned by autoradiography in X-ray films or by PhosphorImager. The mean terminal restriction fragment (TRF) length is determined as the weighted average of the phosphometric signals.

Notes

1. It is recommended to aliquot the 2X HBS solution in smaller volumes and store at -20° C. The pH of this solution is critical for efficient transfection of the plasmids.
2. The PD level per each passage is calculated from the equation

$$\frac{\left(\frac{Nf}{Ni}\right)}{Ln2} \qquad \text{...(1)}$$

   where *Nf* is the number of cells harvested after a given passage and the *Ni* the number of cells plated initially. The kinetics of cellular proliferation can be visualized by plotting the cumulative PDs against the days in culture.
3. T25 tissue culture flasks can be coated with rat tail collagen for better cell attachment during primary cell culture establishment.

Collagen solution can be prepared with 2 mL of the stock collagen solution from Collaborative Biomedical Productsin 100 mL of 0.02 *N* glacial acetic acid. The flasks can be coated with 2 mL of the collagen solution for 10 min at 37°C and washed once with 1 mL KGM.

4. Retroviral vectors are made by transfection of the retroviral expression plasmid into a packaging cell line, which contains the packaging genes. In the packaging construct, the gene encoding the virus' native envelope glycoprotein is replaced by heterologous envelope gene. This process, termed pseudotyping, has great experimental advantages, such as alteration of the virus' tropism and stability. Envelope glycoproteins play a major role in determining the tropism of the virus through their interaction with the receptor molecules on the host cells. Pseudotyping the virus with vescicular stomatitis virus glycoprotein (VSV-G) allows for construction of pantropic virus with broad host cell range, enabling virus entry into most mammalian and nonmammalian cells. Furthermore, VSV-G-pseudotyped virus attains increased stability and can thus be concentrated by ultracentrifugation without significant loss of infectivity. This characteristic of VSV-G pseudotyping can be used to circumvent the problem of inefficient virus production and to salvage low-titer virus.
5. As a result of broad host range, VSV-G-pseudotyped retroviruses impose significant biosafety concerns. The current biosafety guidelines established by the National Institutes of Health (NIH) recommend use of retroviral vectors with biosafety level (BSL) 2 containment with BSL-3 practices. Because retroviral vectors are constructed in a packaging cell line which lacks the complete retroviral genome, the occurrence of replication-competent retroviruses is extremely rare. However, viral genome integrates into the host cell chromosome and poses a risk of insertional mutation. Thus, all laboratory precautions set forth by the institutional biosafety committee must be carefully observed.
6. It is often necessary to determine the infectious titer of the virus. The virus stock is serially diluted in the nutrient medium and used to infect the target cells grown in six-well plates as described above. The number of infected cells per well, reflecting the number of infectious virus particle per unit volume, can be determined by exposing the infected culture to the drug selection marker. The ideal titer of virus may vary depending on specific experimental

need. However, low-titer virus can be concentrated by ultracentrifugation if the virus is pseudotyped with VSV-G. It is important to note that the virus titer is greatly influenced by transfection efficiency of the packaging cells. Thus, the transfection condition must be optimized for preparation of high titer virus.

7. In general, the minimal lethal doses for primary human cells are 1 μg/mL for puromycin, 200 μg/mLl for G418, and 10 μg/mL for hygromycin. However, the ideal antibiotic concentration is cell type-dependent and needs to be determined empirically by titration. The selection procedure generally takes up to 5–7 d, although massive cell death should be apparent within 48 h.
8. It is important to note that the effect of exogenous telomerase activity on the replicative life span is cell type-specific. It is possible to immortalize a certain type of cells by *hTERT* expression alone, as it is the case for HDF, or only in combination with other genetic alteration, such as inhibition of the p16^{INK4A}/pRb pathway. Therefore, even after expression of exogenous *hTERT* in cells, it is possible to observe no phenotypic changes in replication kinetics and life span. The p16^{INK4A}/pRb pathway may be experimentally abrogated by overexpression of E7 oncoprotein of "high risk" human papillomavirus (HPV) or the mutant form (R24C) of CDK4.
9. End-labeling of oligonucleotides is typically performed in 20 μL reaction containing the oligonucleotide primer, 1X kinase buffer, and 1 U PNK for 30 min at 37°C. After labeling, the free unincorporated nucleotides can be removed by gel exclusion filtration in G25 microspin columns. For the TRAP assay and TRF hybridization, purification of the labeled oligonucleotides yields cleaner signals with lower background contamination.
10. One of the common problems with any PCR-based assay is DNA template contamination, resulting in false-positive results. Thus, it is good laboratory practice to compartmentalize the space and equipments, e.g., pipettes, microcentrifuge tubes, and pipet tips, for PCR amplification.
11. To determine the mean TRF length from the telomere-specific Southern blotting, the phosphometric scan of the entire length of the gel is divided into smaller segments using the ImageQuant software. It generally gives more reliable readout when the whole length of the gel is cut off at the upper and lower 20% to provide the segmental phosphometric intensity (*PIi*). Then, the mean TRF

length is determined by calculating the weighted average of the length smear using the equation

$$\frac{\sum PI_i}{\sum \frac{PI_i}{L_i}} \qquad \ldots(2)$$

where PI_i denotes the phosphometric radiointensity at position i along the length of the gel between the upper and lower 20% cut off points and L_i the corresponding length of restriction fragments in kilobase pairs (kbp).

Genes and Life Span

The yeast *Saccharomyces cerevisiae* is one of the favorite models used to study aging. It has the advantage of sharing physiological changes with mammalian cells. Because human disease genes often have yeast counterparts, they can be studied efficiently in this organism.

The replicative lifespan of yeast is defined as the number of cell divisions or daughter cells (D-cell) that mother cells (M-cell) produce in their life. During yeast growth, each cell division leaves a circular bud scar on the surface of the M-cell, specifically at the site of division. Thus the age (counted in generations) of an M-cell can be determined simply by counting the number of bud scars on its surface. We believe that a *wheat germ agglutinin* (WGA)-mediated fluorescent marker can be used to visualize bud scars if specificity of the binding satisfies experimental requirements. We therefore developed a new screening system based on enrichment of old M-cells and repeated re-growth, followed by a *bud scar-based sorting* (BSS) step. Using magnetic microbeads of very small size to purify M-cells was critical to protecting the viability of cells from damage caused by binding.

A frequently used strategy to search for genes responsible for aging is to select survivors after exposure of cells to stress. However, the question remains whether such genes are picked up in response to the stress treatment or because of their direct effect on aging. The screening method described here provides an alternative that allows direct hunting of human genes that confer longevity. Our BBS screening system can also be used to screen genes with potential anti-aging functions from various libraries or library combinations of eukaryotes under more natural low-stress growth conditions.

Reactive oxygen species (ROS) are produced mainly in the mitochondrial organelles, and oxidative damage is increased during ageing. All the indications are that long-lived species have lower ROS production rather than elevated defenses. Such studies can be performed more easily in the unicellular yeast *S. cerevisiae*. Therefore, the BBS screening system could be applied in general for screening genes that associate with reduced ROS production.

In conclusion, the BBS process is a high-throughput method that is efficient, sensitive, rapid, user-friendly, and reliable. The isolated genes can then become instrumental in the rational design of drugs and the development of therapies in the field of age-related diseases.

Materials

Strains

The following *S. cerevisiae* strains were used: INVSc-1 (MATa; his3Δ1; leu2; trp1-289; ura3-52); BY4742 (MATα; his3Δ1; leu2Δ0; lys2Δ0; ura3Δ0); and BY4742-derived Δ*fob1* strain (BY4742; Mat α; his3Δ1; leu2Δ0; lys2A0; ura3Δ0; YDR1 10w::kanMX4).

Yeast culture

1. Rich YPD (standard yeast complete medium) is made by autoclaving for 25 min a solution of 1% yeast extract, 2% bactopeptone and 2% dextrose in H_2O.
2. Minimal medium, also known as synthetic medium (SD-medium), is made by autoclaving for 25 min a solution of 0.67% yeast nitrogen base without amino acids, 2% dextrose and 0.079% complete supplement amino acid mix (CSM) in H_2O.
3. Media for YPD and SD plates are made by adding 2% agar to the liquid media before autoclaving. After autoclaving, flasks are left for 45 to 60 min at room temperature until cooled to 50°C to 60°C, or kept in an oven at 50°C to 60°C. The medium is poured into plates (25 ml/plate). Plates can be stored for up to 1 wk at 4°C.
4. Yeast cells are inoculated in liquid medium and grown at 30°C and 250 rpm. Culture density is monitored by measuring OD_{600}. Yeast colonies are observed on the plate after incubation at 30°C for 3 d.

Cell and bud-scar staining

1. Sulfo-NHS-LC-Biotin from Pierce Chemical Company.
2. WGA conjugated with fluorescein isothiocyanate (WGA-FITC) from Sigma.
3. Streptavidin-R-phycoerythrin (streptavidin-PE) from Molecular Probes (Eugene, OR).
4. Anti-biotin microbead from Miltenyi Biotec.

Flow cytometry

Fluorescence-activated cell sorting (FACS) is performed on a flow cytometer. FITC and PE staining are analyzed at an excitation wavelength of 488 nm, using a 15-mW argon ion laser. FITC is measured as a green signal by the FL1 detector through a 525 nm band-pass filter and PE as an orange signal by the FL2 detector through a 575 nm band-pass filter. For multi-color staining, electronic compensation is applied to the fluorescence channels to remove residual spectral overlap. At least 10,000 events are detected on each sample. Analysis of multivariate data is performed with CELLQuest software.

Method

Use of yeast as an ageing model requires development of procedures for the isolation of an enriched collection of older cells. The BSS system involves cycles of enrichment and re-growth of M-cells, followed by a final sorting to select those with a longer lifespan.

Labeling of M-cells with biotin

D-cells do not have detectable remnants of M-cell walls, so that labeled M-cell wall is not transferred to later generations. Therefore, covalent binding of biotin to the primary amines on the cell wall hallmarks the M-cells throughout the entire period of growth, and facilitates separation from the D-cell population.

1. Cells are grown at 30°C overnight in 5 mL YPD or SD-medium. The next morning, cells are resuspended to OD_{600} of about 0.02 in 50 mL medium and cultured at 30°C and 250 rpm. Cell density should not be allowed to exceed an OD_{600} of 1.
2. Cells are collected by centrifugation for 10 min at 5000 rpm and 4°C, and washed twice with 1 mL of sterile phosphate-buffered saline (PBS; pH 7.2). All cells are harvested and used as initial M-cells.
3. Before labeling, M-cells are washed twice with PBS, resuspended in PBS at a density of 2.5×10^7 cells/mL, and then incubated with 0.2 mg/mL Sulfo-NHS-LCBiotin for 30 min at room temperature with gentle shaking. Free biotin reagent is removed by washing twice with PBS.
4. Biotinylated M-cells are inoculated in liquid medium to OD_{600} approx 0.01–0.02, and then cultured at 30°C for the desired number of generations (up to seven generations in our experiments; culture is not allowed to exceed OD_{600} = 1).

Purification and regrowth of M-cells

The separation of M-cells from their D-cells is carried out by magnetic cell sorting using a MACS separator.

1. 2.5×10^7 biotinylated M-cells are cultured up to seven generations. The cells are collected and washed twice with cold PBS and resuspended in 30 mL cold PBS. They are then combined with 60 μL of anti-biotin microbeads and incubated for 1 h at 4°C. Unbound beads are removed by washing twice with cold PBS. Cells are then resuspended in 5 mL of cold PBS to a density of $<2 \times 10^8$/mL.
2. The LS column is set up on a MidiMACS separator and rinsed twice with 3 mL of cold PBS. Cell suspension (5 mL, $\leq 2 \times 10^9$ cells) is loaded onto the column. The column is washed two to three times with 5 mL of cold PBS to remove nonbiotinylated D-cells.
3. The column is removed from the MACS separator, and the biotinylated M-cells are eluted with 8 mL cold PBS. These M-cells are ready for re-growth or labeling of bud scars. The M-cells after re-growth can be isolated again by the MACS separator system.
4. The purity of M-cells sorted by MACS is examined by staining with fluorescent marker, utilizing the high-affinity specific binding between streptavidin and biotin. About 10^7 biotinylated M-cells are stained with 3 μL streptavidin-conjugated R-phychoerthrin (PE) in 1 mL of PBS for 1 h at room temperature in darkness. The cells are then washed twice with PBS and suspended in 2 mL of PBS. These cells are ready for FACS analysis. In our experiments, contaminating D-cells accounted for less than 4% of the population.

Sorting of old M-cells based on bud-scar staining

A typical signature of ageing in a budding yeast cell is the accumulation of chitin-rich bud-scar rings on the cell surface. To visualize bud scars, yeast cells are often stained with a high concentration of Calcofluor white M2R. However, this reagent is not specific for bud scars because it binds not only chitin, but also the rest of the cell wall. Moreover, high concentrations of Calcofluor can affect cell viability. Based on the ability of WGA to bind with high specificity to chitin polymers, and its inability to penetrate into the cell wall because of its high molecular weight, WGA conjugated with fluorescent dye is used as a marker for specific labeling of bud scars. Bud scars staining with WGA-FITC and biotin-specific staining with

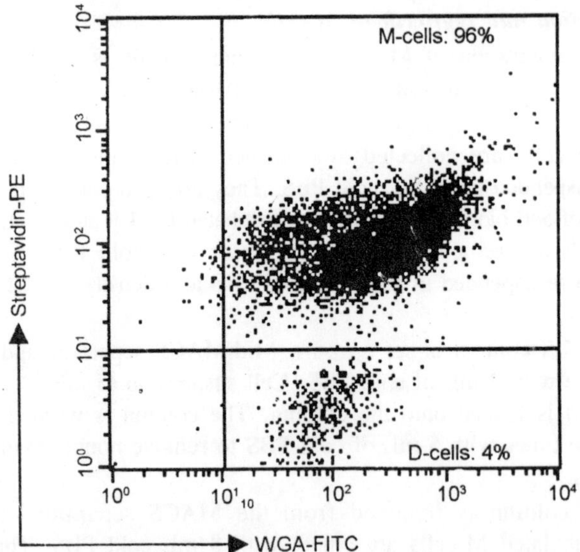

Fig. 3.2. The purity of M-cells after magnetic sorting. 2.5 × 10^7 exponential INVSc-1 M-cells were biotinylated and cultured to seven generations in SD medium.

streptavidin conjugated with R-PE are applied simultaneously to distinguish M-cells from contaminating D-cells. M-cells are double-stained and recognized according to two-color fluorescence (FITC green and PE orange) by the flow cytometer, whereas D-cells are stained only with WGA-FITC.

1. 1 × 10^7 M-cells are incubated with 12 μL of WGA-FITC (1 mg/mL) and 3 μL of streptavidin-PE in 1 mL PBS for 1 h at room temperature in darkness. Cells are then washed twice with PBS to remove free label and resuspended in 2 mL of PBS at a concentration of 0.5 × 10^7 cells/mL.
2. Unstained cells and single stained cells are used for setting parameters. Run the cell suspension through the flow cytometer, observing the forward and right angle light-scatter dot plots (log/log). Adjust the gain setting on the forward light-scatter (FSC-H) and right angle light-scatter (SSC-H) to set the yeast cell population just on the centre of the screen.
3. With unstained cells flowing through the instrument, display FL1-H vs FL2-H dot plots (log/log). Increase (or decrease) the high voltage on the FL1 and FL2 so that the unstained cells are in the lower left-hand corner.

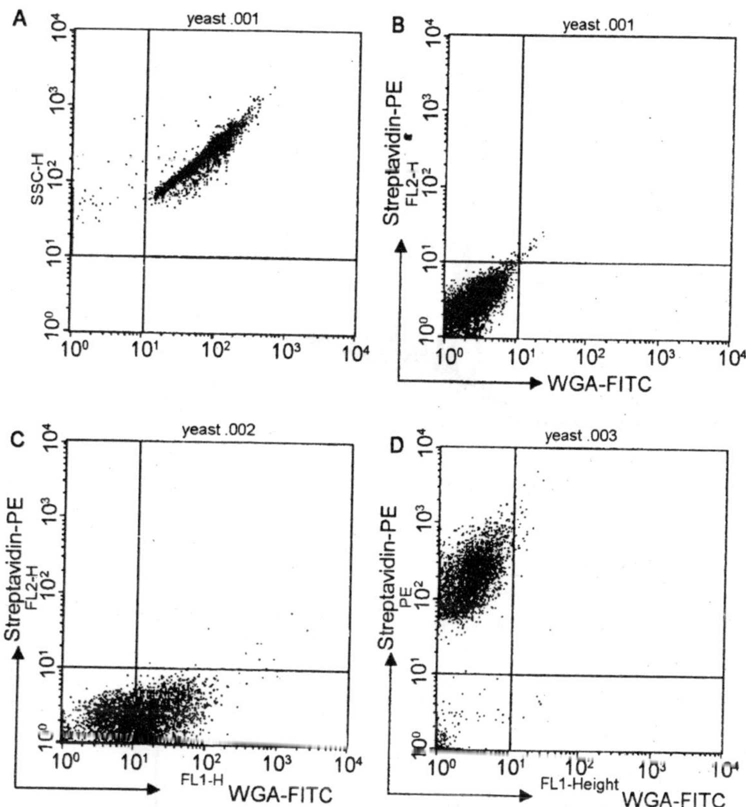

Fig. 3.3. Fluorescence-activated cell sorting analysis of yeast cells.

4. Test cells stained only with FITC and PE for any spectral compensation between these two channels.
5. Double-stained M-cells are recognized by FACS analysis. The fluorescent dye (FITC) is confined to the bud-scar rings, and is hardly detectable on other cell wall parts. The intensity of the fluorescence signal detected by flow cytometry correlates with the number of bud scars on each individual cell.
6. M-cells with a high intensity of FITC signals represent a population with a larger number of bud scars, which allows the sorter to separate and collect cells with the desired number of bud scars. The gate setting for collecting older M-cells is based on two parameters: the fluorescence intensity of PE, which differentiates the M-cells from D-cells, and that of FITC, which helps to select cells with high bud scar numbers.

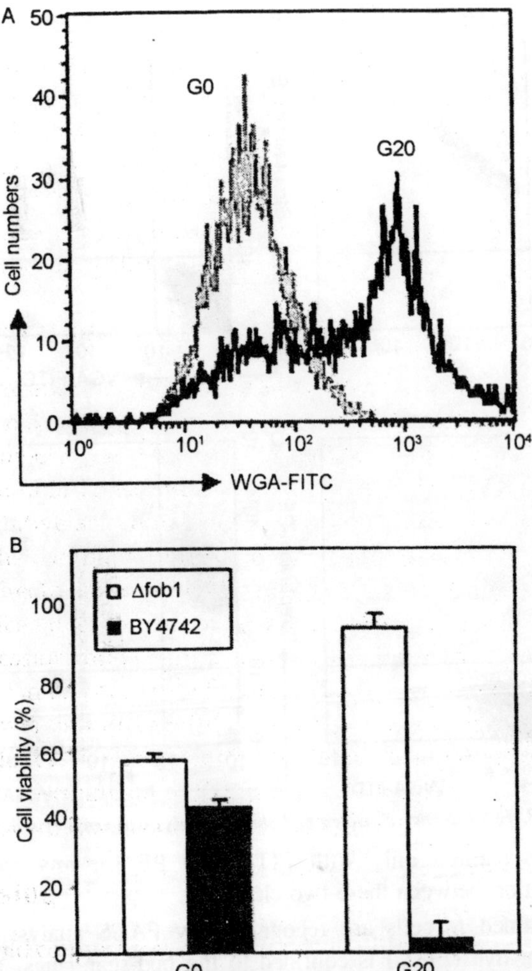

Fig. 3.4. Bud-scar sorting of the yeast Δfob1 strain.

The sorted M-cells are collected and plated on YPD plates. Individual colonies are observed after incubation at 30°C for 2 to 3 d. These colonies, which potentially have a longer replicative lifespan, can be stored and studied further.

Notes

1. Preparation of sterile media is essential for genetic manipulation. Media containing glucose, especially YPD, can be overcooked. Avoid a dark brown color by autoclaving no longer than 25 min and removing the medium soon after the pressure is down.

2. Yeast cells grow much more rapidly in YPD medium than in minimal medium. As cell density increases, nutrient supplies diminish and the rate of cell division slows down. At a density of 5–10 × 10^7 cells/mL yeast culture is saturated, and the cells enter the stationary, or G_0 phase. It is important to keep the M cell growth in the exponential or log-phase stage. Aeration intensity, dissolved oxygen and carbon dioxide, temperature and pH can affect yeast growth. The effects of these parameters are generally strain dependent, and so they must be considered for each strain individually. For good aeration, yeast medium should occupy no more than one-fifth of the total flask volume, and growth should be carried out in a shaking incubator.
3. Complicated synchronization procedures for mothers are not required because the difference among the various M-cell generations does not affect the final sorting/collection of M-cells with desired high age.
4. Smeal et al. successfully demonstrated isolation of M-cells based on the biotinylation of initial mothers, but noticed poor viability and low recovery. To overcome this problem, we modified the method in a critical step by using microbeads (50 nm diameter). Because of their small size, microbeads did not affect cell physiology and therefore bead detachment was not necessary. Cells collected by magnetic sorting without bead detachment behaved like the controls in subsequent re-growth in liquid or on solid medium.
5. The cells need to be suspended properly before loading onto the LS column. Clumps and aggregates are removed by a Cell Strainer.
6. Depending on the amount of biotinylated M-cells, a MACS separator and column with a maximum capacity for isolating M-cells should be selected. In this protocol, we use MidiMACS and LS columns that allow positive selection of up to 10^8 cells labeled with MACS Microbeads from a total of up to 2 ×10^9 cells.
7. During subsequent growth the percentage of initially labeled M-cells in the population declines exponentially in every generation. For example, during 20 generations total cell number will increase 2^{20}-fold. To allow M-cells to reach a high number of cell cycles under optimal, non-stationary phase conditions, huge culture volumes would be required, which is practically impossible. Therefore, after an intermediary growth cycle number (usually seven), the

M-cells are collected by magnetic beads conjugated to a biotin-binding moiety, such as streptavidin or antibody. The collected M-cells are then re-grown in fresh medium. Purification of M-cells was achieved by adding anti-biotin magnetic microbeads with a diameter of 50 nm to the cell culture. Sorting of M-cells grown up to stage G7 was regularly performed and was successful in terms of growth rate, sorting efficiency and cell viability.

8. Analytical test FACS runs are necessary for each experiment. After the test run, the gate is defined to collect 100 to 1000 of the original 3×10^7 M-cells. The FACS analysis shows how many events (cells) are counted above a certain fluorescence level. To quantify the dead cell fraction, dead cell staining (such as propidium iodide) could be applied simultaneously.

4

Life Span of *Saccharomyces cerevisiae*

Understanding the aging process and how to block or delay it is one of compelling challenges of modern science. The last decade has seen the identification of several genes implicated in life span regulation and, most importantly, has pointed to a partially conserved nutrient/insulin/insulin-like growth factor (IGF)-I-like pathway, found in organisms ranging from yeast to mammals, as a master regulator of the aging process. In yeast, the downregulation of two nutrient-sensing pathways, the Ras/protein kinase A (PKA) and the Sch9 pathways, increases stress resistance and can extend the chronological life span by up to threefold. Analogous effects are obtained by decreasing the activity of the insulin/IGF-I pathways in worms, flies, and mice. Therefore, *Saccharomyces cerevisiae* is a simple eukaryote that can serve as a valuable model system for aging in mammals. Its use offers several advantages, above all a relatively short life span and the availability of straightforward genetics and genomics techniques.

The chronological life span is different from the well studied replicative life span, which measures the reproductive potential of individual yeast mother cells. Analogously to the life span of higher eukaryotes, the yeast chronological life span is a measure of the length of time a population of yeast organisms remains viable under nondividing conditions. To provide a paradigm that reflects the conditions encountered in natural environments, the chronological life span is measured for cells grown in glucose-containing medium and maintained in this medium in a postdiauxic (an initial high metabolism phase that follows the growth phase) and stationary phase (a following low metabolism phase) until they either lose the ability to divide or die. Although this may be viewed as a phase that causes death by starvation, it does not appear to be, because in the late phases of growth, yeast cells store reserve nutrients, which do not seem to become limiting during the lifetime of the population. Survival in the postdiauxic and stationary phases is a condition unavoidably encountered by most microorganisms in the wild.

In fact, the longevity trajectories of laboratory strains incubated in grape extract are remarkably similar to those of the same strains incubated in SDC medium, which we use for the chronological life span. Analogously to the postdiauxic cells, stationary phase cells are characterized by high resistance to cellular stress and the accumulation of reserve carbohydrates. As mentioned previously, stationary phase cells' metabolism is low. An even lower metabolism and very long-lived phase, the dormant spore phase, is instead entered by only a small percentage of diploid cells. Thus, depending on the conditions yeast can survive and age in at least three different phases: (1) the high-metabolism postdiauxic phase, (2) the low-metabolism stationary phase, (3) the very low-metabolism spore. In yeast, all three phases are affected by the Ras/PKA, Tor, and Sch9 pathways which appear to control both common and distinct sets of genes to regulate chronological longevity (Ras/PKA/Tor). A similar long-term survival strategy in response to extreme starvation called dauer larva has evolved in the round worm *Caenorhabditis elegans* and is controlled in part by the pathway that regulates longevity (insulin/IGF-I-like). Therefore, it is possible to hypothesize that life span-regulatory mechanisms have originally evolved in microorganisms in order to postpone reproduction in response to starvation.

Experimentally, stationary phase entry is usually induced by growing a yeast population in rich medium (YPD) until saturation and maintaining the cells in the same medium without replenishing any of the nutrients. Under these conditions, yeast grow rapidly on glucose, undergo the diauxic shift when they stop growing and switch from fermentative to respiratory metabolism, resume cell growth at a very low rate, and finally undergo cell cycle arrest and enter the low metabolism stationary phase. Normally, entry into stationary phase occurs 3 to 6 d after the diauxic shift and yeast populations may survive for up to 3 mo before cell viability starts decreasing. Whereas YPD is commonly used for chronological life span measurements, we

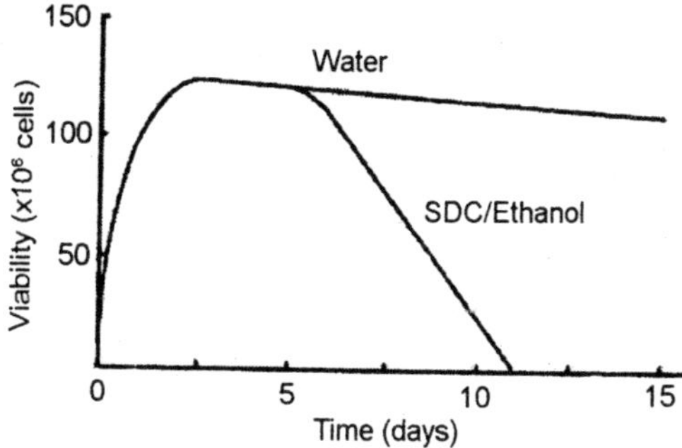

Fig. 4.1. Chronological life span measurement.

noticed that this rich medium induces large fluctuations in cell viability, which may depend on the fact that over time fractions of the arrested yeast population are stimulated to use the nutrients released by the dead yeast to reenter the cell cycle. Therefore, to monitor chronological life span of yeast we mostly use synthetic complete medium (SDC), which appears to promote an easily detectable cellular regrowth only after the majority of the population has died.

In several genetic backgrounds incubation in SDC promotes a shorter but mostly high-metabolism life span (5-6 d) (*postdiauxic phase*) after which yeast die rapidly. Alternatively, we perform chronological survival experiments in water (severe *caloric restriction* [CR]) by switching the cells from SDC to water after the population reaches saturation. Water incubation mimics extreme starvation conditions, which are commonly encountered by yeast (i.e., when they are washed out from fruit by rainwater).

Analogously to yeast incubated in YPD, yeast switched to water enter stationary phase as shown by their reduced metabolism and survive much longer than when maintained in SDC. This extreme form of CR should serve as a powerful model to understand the mechanisms responsible for the well studied effects of CR on the extension of the life span of all model organisms. Importantly, mutations that extend life span in SDC also show the same effect in water, suggesting that the downregulation of either the Ras or Sch9 pathway causes life span extending changes that only partially overlap with those caused by CR.

Materials

1. SDC liquid medium: 0.18% yeast nitrogen base w/o amino acids and ammonium sulfate, 0.5% ammonium sulfate, 0.14% NaH_2PO_4, 0.173% complete amino acid mix (discussed later). Dissolve all

components almost completely in water and adjust pH to 6.0 with NaOH. Autoclave and add dextrose to a final concentration of 2% before use.

2. Complete amino acid mix with adenine and uracil: adenine 80 mg/L, 80 mg/L uracil, 80 mg/L tryptophan, 80 mg/L histidine-HCl, 40 mg/L arginine-HCl, 80 mg/L methionine, 40 mg/L tyrosine, 1200 mg/L leucine, 60 mg/L isoleucine, 60 mg/L lysine-HCl, 60 mg/L phenylalanine, 100 mg/L glutamic acid, 100 mg/L aspartic acid, 150 mg/L valine, 200 mg/L threonine, 400 mg/L serine. This is a modification of the original recipe.
3. Yeast extract/peptone solid medium (YPD): 10 g/L bacto-yeast extract, 20 g/L Bacto-peptone, 20 g/L Bacto-agar. Dissolve in water. Autoclave and add dextrose to a final concentration of 2%. Mix well and pour plates.
4. 20% (w/v) glucose stock solution.
5. Autoclaved 18 MΩ Milli-Ω grade water is used for all media preparations, experiments in water, dilutions for CFUs assay.
6. 100 × 15 mm Petri dishes (VWR).

Methods

Chronological Life Span in SDC/Ethanol (Postdiauxic Phase)

1. Streak a small portion of the frozen stock onto a YPD plate and incubate at 30°C for 2–3 d.
2. Inoculate a few colonies (four to five) into 1-2 mL of SDC medium and grow overnight. Dilute the overnight culture to an initial density of 1-2 10^6 cells/mL (OD600 of 0.1-0.2) in 10–50 mL of SDC. Incubate at 30°C in flasks with a volume/medium ratio of 5:1, shaking at 220 rpm.
3. After 3 d, start monitoring survival by measuring the ability of an individual yeast cell/organism to form a colony (*colony forming units* [CFUs]). Normally, cultures are serially diluted to reach a 1:10^4 dilution in sterile distilled water and 10 μL of the same dilution are plated in duplicate onto two halves of a YPD plate. Particular care must be taken to avoid to spread the diluted culture over the line that marks the separation between the two halves of a plate. Two dilutions per culture are used routinely to reduce the fluctuations due to the manual error. CFUs are counted after 2-3 d. The number of CFUs at day 3 is considered to be the initial survival (100% survival) and is used to determine the age-dependent percent survival.

4. Monitor CFUs every 48 h by adjusting the dilution factor and the volume plated according to the mortality rate. Continue until survival reaches 1% to record a time value that approximates the maximum survival time.

Chronological Life Span Under CR: Incubation in Water, Stationary Phase

1. Proceed as above, steps 1-3.
2. Harvest cells by centrifuging at 1400 RCF for 5 min at room temperature in sterile polypropylene tubes. Resuspend pellet in a volume of sterile distilled water equal to that of the original culture and centrifuge as previously.
3. Wash cells two more times and resuspend them in a volume of sterile distilled water equal to that used for the original SDC culture.
4. Pour the culture into the original flask used for incubation after rinsing it with sterile water.
5. Repeat steps 2-4 every 48 h.
6. Monitor CFUs as described above every 48 h.

Notes

1. Ideally, viability experiments should be started using yeast that have been recovered from frozen stocks and grown on YPD plates for 2-3 d only. Incubation of the master plates at 4°C, especially for an extended period of time (>1 wk), might affect survival significantly.
2. We normally start our experiments by inoculating a few colonies instead of a single one to reduce the effect of stochastic components on survival rates.

   The selection of the size of the flasks used for longevity studies performed in SDC depends on the life expectancy of the strains used. Bigger flasks (250 mL) are recommended for long-term experiments (>3 wk) to avoid medium evaporation.
3. Experiments are normally performed at 30°C. Incubation at higher temperatures (37°C) causes cellular stress and dramatically reduces yeast survival.
4. To test whether the loss of CFUs is an appropriate system to monitor loss of viability, we measured the concentration of proteins released into the medium by dead and damaged wild type DBY746 and long-lived mutants and found an inverse correlation between proteins released and number of CFUs.

5. Older cells tend to reenter the cell cycle more slowly than young ones. At the last stages of a survival experiment incubate the YPD plates for an additional day at 30°C to avoid the underestimation of viability.
6. The number of CFUs at day 3 is selected as initial survival (100%) considering that in our wild-type strains DBY746 and SP1 the population density does not normally increase after day 3, suggesting that the great majority of the cells has stopped dividing.

   When working with strains or mutants that are particularly short-lived, it is recommended to monitor survival daily starting at day 1.
7. We have shown that after more than 99% of wild-type DBY746 and SP1 yeast incubated in SDC dies, in about 50% of the studies a better-adapted subpopulation is able to grow back by utilizing the nutrients released by dead cells. Under these circumstances, regrowth is very easily detected because survival can increase 10- to 100-fold. However, the possibility that a small fraction of cells in the culture restarts dividing before survival reaches 1% cannot be ruled out. If this event were to occur during the early stages of the experiment when the majority of the population is still viable, it might be hard to distinguish it from the "normal" survival fluctuations and it would be likely to interfere with the outcome of the experiment. When we suspect that regrowth might be occurring, we turn to the water paradigm. In fact, incubation in water and the removal of nutrients released by dead organisms achieved by three washing steps miminize the chance of growth during long-term survival.
8. In some genetic backgrounds (i.e., BY4741) survival rate usually levels off after reaching approx 5-10%. It is not clear whether this is due to the selection of extremely long-lived subpopulations or, most likely, to a phenomenon similar to the gasping observed in *Escherichia coli* "stationary" cultures, which is characterized by the regrowth of a fraction of the population capable of utilizing the nutrients released by dead organisms.
9. For survival experiments performed in water, we recommend starting with day-3 SDC cultures that have reached an OD_{600} close to the average one they normally reach (14-15 for DBY746 and SP1). We have noticed that survival in water performed on yeast populations that had not grown optimally in SDC was significantly reduced.

10. We have observed that whereas the manipulation of yeast incubated in SDC is relatively simple and chances of contamination very low, with water cultures the risk of contamination is very high. This depends in part on the washing steps that cause more manipulation of the cultures but also on the fact that the low pH of the SDC cultures reached after 3 d (3-3.5) protects them from contaminants, very likely by inhibiting their growth or killing them. To reduce the risk of contaminations, we recommend the use of sterilized water contained in bottles never opened each time the washing procedure is carried over.

Life Span of Yeast Cells

Measuring the life span of long-lived organisms such as humans, and mammals in general, is an extremely time-consuming task. As a result, researchers have turned to simpler organisms, with shorter life spans, to understand the aging process and to identify genes and small molecules that might regulate it.

Extending the life span of mammals is achievable. A simple dietary manipulation known as *caloric restriction* (CR) can extend life span of rodents approx 40%. The question has been: what underlies this effect? In recent years, the budding yeast, *Saccharomyces cerevisiae* has become one of the leading models for understanding how CR works. Today, more than 20 *longevity assurance genes* (LAGs) have been described for yeast, and we have a good understanding of how CR works, the best for any studied species.

There are two ways to define yeast life span. One is the number of days a yeast culture remains viable in stationary phase. This is known as the "*chronological life span*". Alternatively, one can determine life span for a single cell based on the number of times it divides, known as "*replicative life span*". The cause of aging in yeast, within the context of replicative life span, is the formation of extrachromosomal rDNA circles (ERCs). As yeast divide, the highly repetitive rDNA locus is prone to recombination events which lead to the excision of circular DNA containing origins of replication. This episomal DNA, containing origins of replication, can then autonomously replicate to levels that far exceed the cell's genomic DNA.

The eventual death of the mother cell is presumed to occur through a titration of essential factors. Also, ERCs segregate asymmetrically, being confined to the mother cell, which thereby allows the daughter cells to escape the same fate. Although yeast is a simple eukaryote, determining its replicative life span is not a simple task. It requires the researcher to be adept with a microscope and a micromanipulator, and to have considerable patience. Only with time will one become skilled at this technique, and often beginners do not obtain reliable data consistently the first time around. For all procedures described in this chapter, the strain W303AR5 MATa *ade2-1, leu2-3, 112, can1-100, trp1-1, ura3-52, his3-11, 15, RDN1::ADE2* is referred to. This strain has the *ADE2* gene integrated at the 25s rDNA locus, allowing it to be also used for recombination analysis. It also responds well to CR, achieving an approx 20% extension of life span. A list of other strains used in replicative life span analysis, as well as their published mean and maximal replicative life spans, can be found in the review by Bitterman et al.

Materials

Preparation of yeast for replicative life span analysis

1. 2.2X supplemented yeast extract/peptone (YEP) media: 22.2 g/L Bacto-yeast extract, 44.4 g/L Bacto-peptone in water and filter sterilized with a 0.22-μm vacuum filter. Supplement the media by adding 40 mL/L 0.5% adenine solution and 20 mL/L for the remaining 1% amino acid solutions.
2. 4% agar: Bacto-agar: 8 g of agar is added into 200 mL of water and autoclaved.
3. 18 MΩ Milli-Q grade water is used for all media preparations.
4. Petri-plates: 100 × 15 mm.
5. Amino acid and adenine stock solutions (w/v): 0.5% adenine, 1% uridine, 1% tryptophan, 1% histidine, and 1% leucine solutions are prepared with water and filter sterilized using a 0.22-μm vacuum filter. These supplement the auxotrophies that are particular to the W303AR strain of yeast. Different amino acids may be required for other strains.
6. 1.05X YEP: 10.5 g/L Bacto-yeast extract, 21.1 g/L Bacto-peptone in water and autoclaved. Supplement after autoclaving as with the preparation of 2.2X YEP media, but use half the amount of solutions.
7. 40% (w/v) glucose stock solution.
8. Dissection needle kit.

Magnetic sorting of old yeast

1. Phosphate-buffered saline (PBS): a 10X stock solution is prepared with 1.37 *M* NaCl 27 m*M* KCl 43 m*M* Na_2HPO_4 14 m*M* KH_2PO_4, adjusted to pH 7.4 with HCl if required and then promptly

autoclaved. A 1X working solution is made by diluting 1:10 with sterile water.

2. EZ-link biotin: sulfo-NHS-LC-biotin.
3. Streptavidin coated paramagnetic iron beads.
4. Calcofluor fluorescent brightener 28.

Isolation of ERCs

1. 1X TE buffer: 10 m*M* Tris HCl, pH 8.0 and 1 m*M* EDTA, pH 8.0.
2. Sorbitol solution: 0.9 *M* sorbitol, 0.1 *M* Tris HCl, pH 8.0, 0.1 *M* EDTA are prepared with water and filter sterilized using a 0.22-μm filter.
3. Zymolyase solution: dissolve 0.3 mg/mL Zymolyase (20,000 U/g; ICN Immunobiologicals) in sorbitol solution and store at 4°C.
4. 2-mercaptoethanol.
5. 10% (w/v) sodium dodecyl sulfate (SDS).
6. 5 *M* Potassium acetate solution (do not adjust pH).

Analysis of ERCs

1. Random primed DNA labeling kit.
2. ProbeQuant sephadex G-50 columns.
3. $dCTP^{32}$.
4. Pre-hybridization buffer: 100 μL boiled salmon sperm DNA (5 mg/mL stock ssDNA) 1g dextran sulfate 1 mL 10% SDS 0.56 gs NaCl in 10 mL water. Heat at 65°C for 30 min to dissolve the salts into solution.
5. 10X TBE electrophoresis stock buffer: 0.89 *M* Tris base 0.89 *M* boric acid 20 m*M* EDTA, pH 8.0. A 1X working solution is made by diluting 1:10 with water.
6. 0.4 *M* NaOH solution.
7. 0.25 *M* HCl solution.
8. 20X SSC stock solution: 3 *M* NaCl 0.3 *M* Na_3 citrate. Adjust the pH to 7.0 with 1 *M* HCl. A 2X working solution is made by diluting 1:10 with water.
9. Nylon membrane.
10. Wash buffers: 2X SSC 0.1% SDS (low stringency) at room temp, 1X SSC 0.1% SDS (medium stringency) at 60°C, 0.1X SSC 0.5% SDS (high stringency) at 60°C, 0. 1X SSC (brief wash) at room temperature.

Recombination analysis

1. 1.05X YEP: Prepared in the same manner as previously described.
2. 2.2X YEP: Unlike 2.2X YEP that is prepared for life span assay plates, do not supplement this stock solution with adenine or histidine. The solution can also be autoclaved rather than filter sterilized. Otherwise, it is prepared in the same manner.
3. 3% agar: Bacto-agar: 6 g of agar is added to 200 mL of water and autoclaved.
4. Large Petri-plates: 150 × 15 mm.

Methods

Preparation of yeast for replicative life span analysis

1. The day prior to starting the experiment, the yeast should be streaked onto agar plates containing media identical to that used for the life span assay. We typically use plates containing a 2% final concentration of glucose for normal propagation of yeast. For CR analysis, the plates are made identically, using the same stock solutions, with a final glucose concentration of 0.5%. The yeast should be as fresh as possible. If taken directly from the freezer, allow at least 2 d of consecutive streaking for recovery. Use yeast only from a plate that has been struck within 24 h, i.e., as fresh and actively growing as possible.
2. Using a sterile toothpick, transfer a tiny amount (barely visible to the eye) from the previous night's streak onto a fresh plate that will be used for the life span analysis. The yeast should be a barely visible dot on the plate. Allow the transferred cells to grow for an additional 3–5 h at 30°C.
3. After attaching the inverted plate to the stage of the microscope, use the tip of the fiber optic needle to drag a bunch of cells, i.e., however many will fit under the needle, approx ≥1 cm away from the plated dot of cells. You may need to repeat this several times to move enough cells.
4. Find cells that are relatively small and healthy looking, i.e., round and exhibiting no aberrant morphology. Use the tip of the needle to array 40 cells, in sets of 5, planting a single column. Allow enough spacing between each cell so as to see no more than three to five cells within the field of view using a 10× objective. Each group of five cells can be marked by stabbing the needle into the agar, i.e., use one stab, two stabs, three, four, a right angle, etc., any pattern that will help you to find and keep track of the

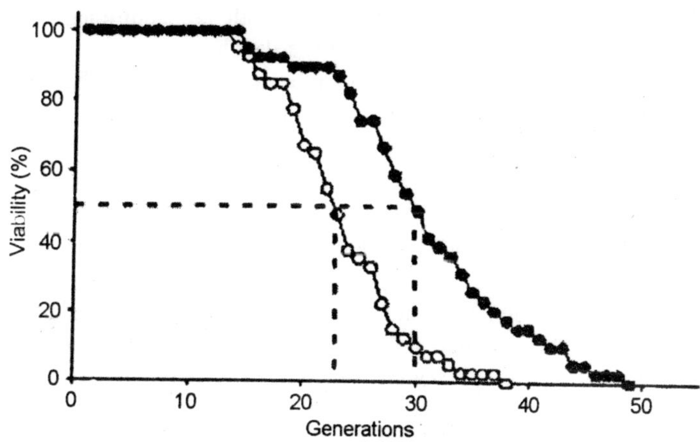

Fig. 4.2. For the replicative life span assay, individual yeast cells are planted in a single column on agar plates, using a 25 μm or 50-μm fiber-optic needle.

column of cells on the plate. The same pattern can then be marked in a notebook with columns divided into groups of five.

5. After arraying the cells, remove the prepared life span assay plate from the microscope stage and incubate at 30°C for 1–2 h, after first sealing the plate with ParaFilm to prevent the loss of moisture. After returning the plate to the microscope, the yeast will have undergone about one to two divisions.
6. With the needle, pull apart the cells and leave behind the smaller budded cell. This starting "*virgin*" cell will henceforth be the mother cell during the assay. Discard the rest of the divided cells by moving them to the left, corresponding to about one-third of a field of view away from the virgin mother cell. Not all cells will be ready for separation at the same time. Thus, you may need to come back after another 30 min to 1 h after the others were separated.
7. Incubate the sealed plate at 30°C for 1–2 h. After transferring the plate back to the microscope, pick off and, from now onward, discard the smaller daughter buds from your starting virgin mother cell and record the number of divisions for each arrayed mother cell. Only count the number of daughter cells that you are able to physically detach from the mother cell, ignoring those that still remain attached until the next micromanipulation session. Note that yeast divide axially so that after one division, one daughter will be attached to the mother cell with either one or no buds forming at the daughter cell. After two divisions, up to four cells

may be picked from the mother cell, with the additional cell corresponding to budded cells from the daughter. Because of the rate of exponential division, it can been readily seen that the formation of 4 daughter cells can lead to up to 16 cells being discarded from the vicinity of the mother. Therefore, it is not advisable to go beyond four divisions, otherwise the mother cell may be hard to pick out. Also, too many cells in one location may lead to excessive nutrient depletion.

8. Repeat step 7 throughout the day. The mother cells become enlarged after progressive divisions and eventually become much easier to discern from the daughter cells. As mother cells age, their rate of division will also slow, with some old cells taking up to 4 h to divide. If a cell does not divide after 8 h, it is considered dead.
9. When all of the cells have died (~10–14 d), their replicative life spans can be graphed. Each row pertaining to each specific cell's division number is tallied and the numbers are plotted as the % cells that are still alive during each round of division. For example, at the start of the assay, 100% of the cells are alive after 0 divisions have occurred, whereas 0% of the cells are alive at the amount of divisions that the last remaining mother cell has undergone.

Magnetic sorting of old yeast cells

1. Inoculate 5 mL of 1X YPD with yeast and grow overnight at 30°C while shaking at 200 rpm.
2. The next morning, resuspend enough cells into 50 mL 1X YPD to reach an optical density at 600 nm (*optimal density* [OD] 600) of approx 1.0 (2×10^7 haploid cells/mL) later that day. This usually entails diluting approx 1 mL of overnight culture into 50 mL 1X YPD. The starting OD 600 should be approx 0.15–0.20.
3. When the OD 600 reaches approx 1.0, spin down cells with a clinical centrifuge set to 4000 rpm for 5 min. Wash cells once in 1 mL of sterile 1X PBS. Resuspend the cells in 1 mL 1X PBS and transfer to a microfuge tube.
4. The cells are now ready to be biotinylated. Remove the EZ-link biotin from the freezer and thaw at room temperature. It is VERY sensitive to moisture. When thawed, add 8 mg of sterile EZ-link biotin per 1×10^8 cells. Gently shake the cells for 15 min at room temperature.
5. Wash the cells seven times using 1 mL 1X PBS to remove the excess biotin label.

6. Resuspend cells in 1 mL YPD. Determine the cell density using a hemacytometer. We use a Bright-Line hemacytometer. Add 2 $\times$ 10^8 cells to 4 L freshly made 1X YPD (autoclaved for 20 min). Shake overnight at 30°C.
7. Spin down the cells at 5000 rpm and thoroughly aspirate off the 1X YPD medium. Resuspend the cells in 20 mL cold 1X PBS and place on ice.
8. Wash the magnetic beads twice with 1X PBS. Add 250 μL of the washed magnetic bead slurry per 1 $\times$ 10^8 cells. Incubate on ice with occasional swirling for 2 h in order for the biotinylated cells to bind to the magnetic beads. The sorting procedure must be performed in a 4°C room. Add equal volumes of the resuspended magnetic bead-coated cells into 16 $\times$ 150 mm glass culture tubes and insert into a magnetic tube rack. We use a BioMag tube rack for the magnetic separation. Wait 15–20 min for the magnetic bead-coated older cells to move towards the side of the tube facing the magnet. Remove young cells that are still in solution gently with a 10-mL pipet and save the solution. Add the same volume of cold 1X YPD into each tube so that the meniscus reaches the top of the magnet and wait 15–20 min again for magnetic separation to occur. Repeat step 6, using cold 1X YPD each time and vortexing gently to resuspend the bead-coated cells. Repeat this step eight times. Check the yield using a hemacytometer. The recovery is typically 50% of the starting amount of cells.
9. Count bud scars by first adding 20 μL of cells into 100 μL 1X PBS. Add a few grains of calcofluor fluorescence brightner and incubate for 5 min at room temperature. Add 1 mL 1X PBS and centrifuge at 6000 rpm in a microfuge for 20 s. Wash cells once in 1 mL 1X PBS, spin down again and resuspend in 100 μL of 1X PBS. With a single sorting most cells will have four to nine bud scars.
10. Place approx 10 μL of cells onto a glass slide and cover with a coverslip. Observe the stained bud scars using ultraviolet (UV) fluorescence microscopy. If cell sorting has been successful, ther old population of cells should contain on average 5–12 more bud scars per cell than the young population of cells.

Isolation of ERCs

1. Resuspend cells in 500 μL sorbitol solution. Add 25 μL zymolyase and 50 μL of 0.28 *M* 2-mercaptoethanol (20 μL 2-mercaptoethanol in 980 μL water). Depending on the activity of the zymolyase,

incubate at 30°C for 15–45 min. To monitor the efficiency of spheroplasting, i.e., digestion of the cell wall, add approx 2 μL of 10% SDS to a 10 μL aliquot of cells. If 85–95% cells appear lysed when viewed under a microscope, then they are ready for the next step.

2. Add 80 μL of 10% SDS, invert several times to mix, and incubate for 20 min at 65°C. The solution should become viscous.
3. Add 200 μL of 5 *M* potassium acetate solution. Invert several times and leave on ice at least 30 min. A white precipitate should form.
4. Centrifuge for 3 min at high speed. Remove supernatant and add 1 mL 100% ethanol. Mix well by inverting several times and centrifuge at max speed for 10 min.
5. Resuspend the DNA pellet in 300 μL TE. Add 1 μL of 10 mg/mL of RNase and incubate for 30 min at 37°C.
6. Precipitate the DNA by adding 1 mL of ice cold 100% ethanol and centrifuging at maximum speed for 10 min. After removing the ethanol, dry the pellet in a speed vac and resuspend in 50 μL of TE. The resuspended DNA pellet can now be stored frozen at -80°C.

Analysis of ERCs

1. The analysis of ERCs essentially involves performing a Southern blot and probing for ERCs with a radiolabeled rDNA fragment. Begin by casting a large 1% agarose gel using 1X TBE. Load samples, along with molecular weight markers and electrophorese overnight at 30 V.
2. Soak the gel for 30 min in 500 mL water containing 25 μL of 10 mg/mL ethidium bromide solution and photograph with a ruler using UV light.
3. Wash the gel for 30 s with water and then soak the gel for 30 min in 0.25 *M* HCl, with gentle rocking.
4. Rinse the gel in water again and soak in 1 L of 0.4 *M* NaOH for 20 min, along with gentle rocking, to denature the DNA.
5. Set up a transfer apparatus, using 0.4 *M* NaOH as the transfer buffer. The transfer apparatus consists of a shallow glass tray that is filled two-thirds of the way to the top with transfer buffer. A glass plate is then used to partially cover the top of the tray, leaving a 1–2 inch gap on either side. A rectangular piece of Whatman 3 MM paper is then cut, wet in transfer buffer, and

placed over the top of the glass plate, with the ends sufficiently long enough to soak in the transfer buffer through the 1–2 inch gaps. This will act as a wick for the transfer. The wick should also be approximately an inch wider, on either side, than the nylon transfer membrane.

6. Drain the gel and place carefully on top of the wick, using a glass pipet to gently roll away any air bubbles that may be trapped under the gel.
7. Cut a piece of nylon transfer membrane, corresponding to the size of the gel. Wet the membrane in transfer buffer and lay on top of the gel, using the pipet to smooth away any air bubbles.
8. Cut three pieces of Whatman 3 MM paper the same shape as the nylon membrane, only a few millimeters smaller at each edge. Soak the piece in transfer buffer and lay each of them on top of the membrane and gel, again using the glass pipet to carefully smooth away any air bubbles after each piece is laid.
9. Place a stack of paper towels, 3–4 inches high, on top the gel set-up, lay a piece of glass plate over the stack, and add a light weight on top to stabilize the plate. Use a bubble level to ensure the plate is perfectly level.
10. Wrap the outer part of the transfer apparatus; i.e., the exposed wick, with plastic wrap to prevent evaporation of the transfer buffer. The transfer apparatus is now left to transfer for 12 h.
11. Rinse the membrane with 2X SSC to remove any agarose residue. The wet membrane is then irradiated with UV light, at an intensity of 0.12 Joules/cm^2. This corresponds to a setting of 1200 using a Stratagene UV crosslinker.
12. Add 10 mL prehybridization buffer to a 200-mL capacity cylindrical glass hybridization bottle. Place the transferred membrane into the bottle and incubate for 1 h at 65°C while rolling.
13. The radiolabeled probe is prepared as per kit instructions. After the reaction is complete, bring the probe reaction volume up to 50 μL with 1X TE. Prepare a Sephadex G-50 column by spinning for 1 min at 3000 rpm, in order to compact the bead matrix. Pipet the 50 μL probe reaction volume into the prepared Sephadex G-50 column and spin for 2 min at 3000 rpm.
14. Boil the radiolabeled probe DNA for 5 min and carefully add to the bottle containing your membrane and prehybridization solution. Incubate overnight at 65°C.

15. The membrane is washed with the wash buffers until counts are reduced to only severalfold over background. The exact washing times must be empirically determined by the user. The bound probe can now be analyzed using either a Phophorimager cassette or X-ray film.

Recombination analysis

1. Cells are first inoculated in the appropriate media, i.e., containing 2% glucose for controls and either 0.1% or 0.5% glucose for calorie restricted cells, and pre-incubated overnight at 30°C while shaking. The following day, cells are re-inoculated into fresh media at an OD_{600} of approx 0.15–0.20 and allowed to grow until log phase has been reached (OD_{600} of 0.8–1.0).
2. Prepare serial dilutions of cells in 1X PBS until a final concentration of 3–4 × 10^3 cells/mL is reached. Pipet 500 μL of this final dilution onto large plates containing 2% glucose containing YPD that is not supplemented with adenine and spread evenly using sterile glass beads along with a gentle shaking motion.
3. After the plates are dry, incubate at 30°C until colonies are large enough to be easily counted by eye. This usually takes about 2 d of growth at 30°C. Plates are then transferred into 4°C for at least 2–3 d or until red pigmentation becomes evident, following a marker loss event.
4. Half-sectored colonies are counted by eye on each plate. Colonies that are completely red are subtracted from the total colony number on each plate. A half-sectored colony represents a marker loss event that has occurred during the first cell division. The wild-type W303AR strain, when grown in media containing 2% glucose, has a recombination frequency of approx 1 × 10^{-3}, whereas deleting *SIR2* elevates the frequency of recombination about 10-fold, to approx 1 × 10^{-2}.

Notes

1. Because the procedure takes approx 10–14 d to complete, it is imperative that the agar plates do not over-dehydrate. Plates should always be carefully wrapped in plastic wrap or Para-Film when in the incubator or cold room. Plates should also be poured with more agar than usual to take into account dehydration and shrinkage of the agar. We usually pour the plates with the agar level reaching two-thirds to three-quarters of the way to the top of the plate.

2. Plates should be poured at least a day before using so that excess moisture is given ample time to evaporate, otherwise dissection becomes difficult.
3. For life span dissections, we have found it best not to autoclave the 2X YEP medium, but to instead filter through a 0.22-μm filter prior to use. We also supplement the medium with additional amino acids, corresponding to the auxotrophies that are unique to a strain.
4. We use Petri plates that have been removed from the plastic wrapper and left out on the bench overnight, with the covers closed. This allows any potentially volatile plasticizers to evaporate from the plates.
5. It is best to drag the cells across the surface of the agar with the needle, rather than lift them off of the surface. This ensures that the cells are continuously in contact with the media.
6. As the plates are continually opened and closed during this protracted time period, it is important to avoid contamination. If a small fungal or bacterial growth occurs far from the plated yeast, it can be successfully removed by cutting away the contaminated section with a sterile scalpel blade. Another troublesome source of contamination may be the presence of fruit flies which, given the chance, will walk all over the surface of the plate, usually ruining the experiment.
7. The plates should also be moved gently because a sudden jarring motion may cause a cell to drop off.
8. Try to dissect the cells everyday. Keeping the plates in the cold room for an entire day is acceptable, but longer periods, i.e., an entire weekend, should be avoided. The rate at which the cells are dividing will determine how many times they can be dissected during the day.
9. Beginners should dissect no more than two columns at once, i.e., one experiment and one control. Later on, as skill level and rapidity increases, three to nine columns of cells may be attempted. We would discourage placing more than three columns of cells on one plate, because this increases nutrient depletion and allows the plate to dehydrate faster, as each individual plate is kept open longer during dissections.
10. The more dissections that are performed per day, especially during the first week, the better the results, i.e., longer overall life spans.

11. Old mother cells are especially fragile, so one should not "*bash*" them with the tip of the dissection needle to dislodge recalcitrant daughter cells. At this point, they have a higher tendency to "*pop*" under pressure.
12. Obviously, mother cells should not be allowed to over-grow because it then becomes next to impossible to keep score, but they should also not be dissected too often because we have found that this tends to shorten their life span, possibly as a result of mechanical damage. Two to three divisions between each dissection session for the W303AR strain is about right.
13. If the agar eventually shrinks to a point at which it is difficult for the dissection needle to reach the surface, part of the plate's edge may be then carefully cut away using a hot scalpel to allow for more freedom of motion. However, it is best to reserve this until absolutely necessary because the larger opening will now permit faster dehydration. Also, when cutting the plate, it should be tilted upwards to prevent any toxic fumes from the molten plastic from coming into contact with the cells.
14. The beads we originally used were purchased from Perceptive Biosystems, which is no longer in business. Thus, in order to perform the magnetic sorting procedure, beads from another source must be substituted. Thus the precise amount of beads used must be empirically determined by the investigator.
15. When the yeast cell wall is digested, the spheroplasts will have a "*ghostly*" or clear outline when viewed under a light microscope. In contrast, yeast with intact cell walls will appear much more refractive.
16. Large plates for the recombination analysis should contain 2% glucose, regardless of the starting glucose concentration during liquid culture, otherwise the red coloration will not develop.
17. It helps to store the plates for several days at 4°C after colonies arise. Over time, the red coloration becomes more intense, allowing for easier detection of half sectored colonies.
18. The omission of histidine from the large plates allows for a more intense red coloration to develop, which allows for easier detection of sectored colonies.

5

LIFE SPAN OF *DROSOPHILA*

Nutrigenomics, which entails analyzing nutrient-gene interactions on a genome-wide scale, was made possible by the sequencing and annotation of numerous genomes. Nutrigenomics is important because improper diets are risk factors for disease. To study diet–gene interactions, whole-genome microarrays can be used to take advantage of the numerous whole-genome sequences that are available. Microarrays have now been developed that can measure the changes in mRNA expression of every gene in the genome when a particular nutrient is added to the diet, for instance. Also, nutrigenomics approaches can be used to identify markers of aging, disease predisposition, and for behavioral genomics.

The fruit fly *Drosophila melanogaster* is ideally suited for conducting many types of nutrigenomic studies. Unlike yeast or *Caenorhabditis elegans*, *Drosophila* have fat bodies with adipocytes, and thus are more similar to mammals. *Drosophila* also have conserved metabolic and signaling pathways involved in fat metabolism, insulin signaling, among many others. Other advantages of *Drosophila* are its sophisticated genetics, small genome size, high fecundity, low cost, and short generation time. Mutations are available in over 50% of the genes and deficiencies have been generated that uncover over 90% of the *Drosophila* genome. More than 76% of human disease genes are conserved in *Drosophila*. Inexpensive DNA microarrays have been developed to analyze gene expression changes cause by altered genotypes and environments. Also, a growing number of insect and other animal species with distinctive ecological niches are being sequenced, opening the door for both comparative genomics and "*adaptive*" nutrigenomics.

The challenge for the future is to use nutrigenomics approaches in *Drosophila* and other model organisms to provide clues to human gene-nutrient interactions.

Unquestionably, it is more informative to present experimental data to illustrate how nutrigenomic studies can be performed in *Drosophila*. Therefore, in this chapter we describe preliminary data on the effects of three diets fed to adult flies on gene expression and longevity. The three diets that we investigate in this chapter are: (1) a high palmitic acid diet (high saturated fat diet), (2) a high soy diet (Asian-like diet), and (3) a high beef diet (Atkins-like diet). Previous studies with these three types of diets in *Drosophila*, humans, and other models are discussed in the remainder of this introductory section.

High-Fat Diets Decrease Longevity in Drosophila

CR or, more precisely, dietary restriction, increases life span in *Drosophila* and many other animals, including humans, and is not an emphasis of this chapter. In contrast, saturated fats in diets that are otherwise isocaloric (identical calorie) have been shown to decrease both mean and maximal life span of *Drosophila*. However, the specific mechanism responsible for the deleterious effects of saturated fats is unknown.

One of the first studies that attempted to make a connection between *Drosophila* dietary components and longevity found that isocaloric diets consisting of high saturated fats (such as palmitic acid) and low carbohydrates will, on average, shorten the life span of *Drosophila* compared with flies fed "control" diets high in carbohydrates and low in saturated fats. These early studies, performed in the late 1970s and early 1980s by Driver and colleagues, did not determine the specific metabolic processes that were negatively affected by the consumption of fat. Instead, they potentially laid the groundwork for further studies.

Unfortunately, however, these early studies by Driver and colleagues were not followed up by other investigators. Other than dietary restriction, very few recent studies have been done on the effects of diet on longevity in *Drosophila*. Likewise, other than dietary restriction, little has been done on the effects of diet on larval development. With the genomic sequences, micro- arrays, mutant lines, and improved technology that are available today but was not available two decades ago, we are now in a better position to perform detailed nutrigenomic studies and understand global effects of dietary fat on health span and lifespan.

Health Benefits of Soy

Scientists are beginning to understand the potential health benefits of soy in our diets. In 1999, the US Food and Drug Administration (FDA) recommended that "25 grams of soy per day, as part of a diet low in saturated fat and cholesterol, may reduce the risk of heart disease". Commercial soy products, such as tofu and soy milk, typically consist of approx 37% protein, approx 37% fat, and approx 26% carbohydrates (% Kcal). Soy protein provides all essential amino acids and is equal in quality to meat and milk protein. Meat and soy combination foods help schools meet the Dietary Guidelines for Americans, and improve the nutritional profile of school meals. The fatty acid composition of soy consists of approx 61% polyunsaturated, approx 24% monounsaturated, and only approx 15% saturated fats. Soy oil is one of the few good sources of omega-3 fatty acids, such as α-linolenic acid, which have been documented to reduce the risks of several chronic diseases, such as arthritis, cancer, and coronary artery disease.

In addition to protein, fat, and carbohydrates in soy, there are also biologically significant amounts of saponins, isoflavones (flavanoids) and other phytochemicals that have been implicated in reducing serum cholesterol and lipid levels. Lucas and colleagues reported that, in Golden Syrian hamsters, soy extracted with ethanol, a treatment which removes saponins, isoflavones, flavanoids, and other phytochemicals, had no cholesterol reducing effects. It is likely that not only one, but several components of soy have synergistic or additive health benefits, reducing cholesterol levels and obesity. Saponins, amphiphilic molecules which are characterized by the property of producing a soapy lather, are a structurally diverse group of triterpene or steroid glycosides that have reduced cholesterol levels in model organisms. Some saponins form insoluble complexes with cholesterol, and when this occurs in the gut, it inhibits the intestinal absorption of both endogenous and exogenous cholesterol.

The possible effects of soy isoflavones, such as daidzein and genistein, on serum lipid and cholesterol levels are less clear. Wagner and colleagues have shown in monkeys that soy protein with isoflavones, but not an isoflavone-rich supplement, improves arterial low-density lipoprotein metabolism and atherogenesis. Yousef and colleagues analyzed the effects of purified isoflavones on seminal (ejaculate) plasma biochemistry in male rabbits. They found that "total cholesterol, percentage cholesterol (out of total lipids), and high density lipoprotein were significantly ($p < 0.05$) increased, while triglyceride did not change in seminal plasma of treated animals". They also analyzed the effects of purified isoflavones on blood serum biochemistry in rabbits. According to the authors, "results showed that isoflavones caused a significant decrease ($P < 0.05$) in the levels of plasma total lipids (TL) by 16% and 19%, total cholesterol by 20% and 20%, triglyceride (TG) by 18% and 23%, low density lipoprotein (LDL) by 19% and 22%, very low density lipoprotein (VLDL) by 18% and 23%, and LDL:HDL ratio by 36% and 39% for 2.5 or 5 mg/kg B.W. doses, respectively, as compared to the control, while the level of high density lipoprotein (HDL) increased by 29% and 32%".

Epidemiological data suggests that soy-rich diets reduce the incidence of prostate cancer in men. Prostate cancer is one of the most commonly diagnosed cancers in the Western world, with nearly a 10% lifetime risk. Boyle and colleagues project that the incidence of prostate cancer will double in the next 30 yr. However, there is a 30-fold greater incidence of prostate cancer in the United States than in the Japanese male population in Osaka, Japan, and a 120-fold higher rate for men in Shanghai, China. Rather than high fat intake being the "root of all dietary evil" and a potential cause of cancer, Griffiths states in a review on soy and cancer that "the elevated plasma levels of phytoestrogens are now seen to be protective, and many believe that estrogens are more implicated in prostate physiology than hitherto thought".

As with prostate cancer in men, epidemiological data suggests that soy-rich diets reduce the incidence of breast cancer in women. Decreased risk of breast cancer has been associated with lifelong consumption of plant foods containing estrogenic compounds, particularly soy isoflavones. Data collected in the Shanghai Breast Cancer Study suggested a protective effect of diets containing isoflavones such as in Asian countries. These studies have generated a great deal of interest in the potential beneficial effects of soy in reducing the effects of cancer. More research on how dietary soy affects cancer development is needed.

Controversial Health Benefits of Low-Carbohydrate and High-Beef Diets

The recent epidemic of obesity in Western countries has contributed to the dramatic increase in the popularity of low-carbohydrate diets, such as the Atkins, South Beach, and Protein Power diets. This rapid switch in dietary habits has had a profound effect on the food industry, despite the opposition from the majority of the nutrition establishment. Recently, some nutrition professionals have argued for the need to understand the apparent success of low-carbohydrate diets in some studies, but little work has been done in this area. Feinman and Fine have argued that low-carbohydrate diets have a "*metabolic advantage*" because proteins require metabolic energy to convert them to carbohydrates.

One component of the inefficiency in metabolizing food is measured by "*thermogenesis*" (thermic effect of feeding), or the heat generated during the metabolism of food. As summarized in a recent review by Jequier, the thermic effects of nutrients is 2–3% for lipids, 6–8% for carbohydrates, and 25–30% for proteins. Feinman and Fine used these values to calculate "*effective yields*" of various diets — a 2000 Kcal diet with the recommended composition of carbohydrate:fat:protein of 55:30:15 has an "effective yield" of 1848 Kcal. However, when the amount of carbohydrates is reduced to 8% of the total Kcal, which is the recommendation of the early phase of the Atkins, South Beach, and Protein Power diets, an additional 140 Kcal are lost as heat per 2000 Kcal base amount. Recently, these "*fad diets*" have been losing their popularity, especially after the recent, highly publicized study that shows that "*overweight*" people (*body mass index* [BMI] between 25 and 30) do not have a significantly higher mortality rate compared with "*normal weight*" people.

Nutrigenomics Research on Dietary Beef

A major component of many low-carbohydrate diets, such as the Atkins diet, is 95% lean ground beef, typified by a "hamburger without the bun". Lean ground beef consists of 35% fat and 65% protein (%Kcal), in addition to 65 mg cholesterol and 55 mg sodium per 85 gram portion. The fat portion consists of approximately equal amounts of saturated and mono-unsaturated fats (FDA food label). Surprisingly, we could only find one previous study in which *Drosophila* were fed dietary beef, and one microarray study where mice were fed beef tallow.

The *Drosophila* beef paper was published in 1979 and the title is "Failure of irradiated beef and ham to induce genetic aberrations in *Drosophila*". The purpose of this paper was to allay the unjustifiable fear, which, unfortunately, is still common among the public, that irradiated food is carcinogenic in humans. In addition to the mouse microarray paper cited previously, several other laboratories also fed mice pure beef fat (tallow) to induce obesity. However, few humans consume large amounts of pure beef tallow, so the application of these results to humans is questionable.

In the experiments described in this chapter, we fed flies a diet containing 95% lean ground beef in order to determine the metabolic effects of consuming beef. Surprisingly, to our knowledge, 95% lean beef diets have not yet been used for nutrigenomics studies in animal models. As described below, our findings suggest that diets rich in 95% lean ground beef, soy, or palmitic acid prolong the larval period, but decrease the mean and maximal life span. However, only beef was able to significantly decrease triglyceride levels in adult flies. In the final section, we will discuss the implications of these findings, if any, in terms of human health and life span.

Materials

Importance of Using Isogenic Drosophila Strains

It is important to use isogenic strains of *Drosophila* that differ in only one gene or genomic region because background strain variation is likely to have a greater affect on life span, metabolism, and gene expression patterns than environmental manipulations. Thousands of *D. melanogaster* strains are available from the main *Drosophila* stock center in Bloomington, Indiana, and in smaller centers throughout the world. However, the most useful collection of PBac transposon insertions and isogenic deficiencies, at least for nutrigenomic studies, were recently released to the *Drosophila* community by Exelixis, Inc. These strains are all derived from a well characterized isogenic strain, *iso-w*; *iso-2*; *iso-3*, from the Bloomington Stock Center. The Exelixis collection includes more than 29,000 PBac transposon insertion lines that tag more than 7200 different genes. The Exelixis isogenic deficiency collection, together with those generated by Kevin Cook and colleagues at Indiana University, also includes 519 deficiencies that uncover approx 56% of the *Drosophila* genome. Because these deficiencies generally uncover relatively small regions of the genome (100,000 to 300,000 base pairs), there is a smaller chance of deleting multiple genes in the same pathway, which was a problem with previously generated deficiencies, most of which were considerably larger.

Dietary Conditions

For normal rearing conditions, we use standard cornmeal-agar-sugar fly food. For specialized food, such as palmitic acid, soy, and beef-containing recipes, we followed a protocol that was derived from Driver and colleagues. The food recipes that we followed from Driver and colleagues contains cornmeal, agar, oatmeal, and either sucrose or an isocaloric amount of palmitic acid, tofu, or beef. Most aging laboratories use sucrose-yeast food (no corn meal or oat meal) for better control of the nutrititve content.

Other dietary considerations are diets that have identical phosphate levels, protein levels, lipid levels, glycemic index, and so on, but vary in the source of these components. We have not yet performed experiments in *Drosophila* with these parameters controlled, but they are important considerations depending on the nutritional question being asked. For example, in mammalian models, intake of a low-phosphate diet stimulates transepithelial transport of inorganic phosphate (Pi) in the small intestine, which is associated with a change in the apical localization of NaPi cotransporters. A low-Pi diet can also lead to an increase in the level of vitamin D3 absorbed in the small intestine in mammals.

The source of protein or lipids, whether from meat or soy, can also potentially affect metabolism, so iso-protein or iso-lipid diets are other considerations. For example, according to Ascencio et al., "The consumption of soy protein was shown to reduce blood lipids in humans and other animal species. Furthermore, it was shown that the ingestion of soy protein maintains normal insulinemia".

Lipids can be considered either "good" (such as polyunsaturated fatty acids) or "bad" (such as saturated fatty acids), so iso-lipid diets are a third consideration. For example, Tatematsu et al. (2005) have shown "Canola oil (Can), as well as some other oils, shortens the survival of SHRSP rats compared with soybean oil (Soy)". It might be worthwhile to determine whether specific protein or lipids have similar effects in *Drosophila*.

Finally, depending on the goals of a particular nutrigenomics study, other considerations must also be made about the food components. One might want to consider glycemic index (GI), which is related to the speed in which nutrients are digested and absorbed, or glycemic load (GL), which describes the total GI content of the diet. Simple sugars have a high GI, whereas complex carbohydrates have a low GI. According to Colombani, in a review on the importance of GI on human nutrition, "It is claimed that low-GI and -GL diets favorably affect many noncommunicable diseases that are prevalent in developed countries, including type II diabetes, insulin resistance, obesity, cardiovascular disease (CVD), and cancer". Although, to our knowledge, it has not yet been investigated, it is likely that GI is also important for regulating metabolic processes such as disease resistance or insulin signaling in *Drosophila*.

Growth Conditions

Climate conditions

In *Drosophila* and other cold-blooded invertebrates, life span is inversely proportional to temperature. Therefore, it is important that relative longevity experiments be performed under constant climate conditions. For the experiments presented in this chapter, flies were reared at 25°C under 60–80% humidity in a climate-controled incubator. To control for circadian alterations, 12 h light: 12 h dark cycles were used, and microarray and metabolic assays were performed with flies isolated at the same time each day (~6 h light).

Larval density

Culture larval density does have a strong effect on body size and likely on lipid content. Therefore, the experiments outlined in this chapter were carried out in culture conditions with controlled larval density. To control for culture density, no more than 50 first instar larvae or eggs were removed and placed in new vials that contained approx 10 mL of food. Virgin males and virgin females were randomly collected on days 10–16 from the vials. For bottles, which contain approx 50 mL of food, 200–300 larvae or eggs were added to each bottle. Vials and bottles also contained folded filter paper or several small sterilized KimWipes, respectively, so that the third instar larvae had sufficient room to crawl and pupate.

Adult culture conditions

For the longevity experiments with mated adults, no more than 50 virgin males and 50 virgin females were added to each cage. A cage consists of a 9-cm Petri dish (Fisher, Inc,) with 10 mL of food and a 100-mL plastic beaker (Fisher) snapped on top. The Petri dishes of food were dried overnight at room temperature with their lids closed before use because adults can stick to wet food. The plastic beaker has several dozen pin holes to allow airflow that were made by heating up a 20-gauge needle with a Bunsen burner and poking the hot needle through the plastic. Every 2 d, the adults were transferred to fresh food in a new cage and the number of dead males and females were counted.

We ensured that the holes in the simple cages were too small for adult flies to climb in our out of, and we never observed contamination of other strains of flies in the cages; nevertheless, in cases where cages, rather than vials or bottles, are required, most laboratories use "*cohort*" cages designed in the Curtsinger, Tatar, and Promislow laboratories. The changing of food and removal of dead animals is much simpler in "*cohort cages*" compared with the beaker and Petri dish system that we used.

Buffers and Preparation of Foods

Homogenization buffer

This buffer is used for triglyceride measurements (0.01 *M* KH_2PO_4, 1 m*M* EDTA pH 7.4). Flies were frozen in liquid nitrogen and ground in this buffer.

Preparation of foods

1. Add 5.4 L of water to a 10-gallon electric steam kettle and boil.
2. Mix in: 500 g corn meal, 250 g oat meal, 50 g dried yeast, 30 g agar.
3. Cook for 20 min.
4. Split into four equal portions in large plastic beakers, labeling them a,b, c, and d.
5. Add: 166.7 g sucrose to beaker a, 66.7 g palmitic acid to beaker b, 366 g Mori-Nu Silken Tofu (extra firm) to beaker c, and 114 g extra lean (95%) ground beef (pre- boiled, dried, and granulated) to beaker d.
6. Mix the sugars into solution with a wooden stick. When the beakers cool down to 60°C, add to each beaker: 5 g methyl 4-hydroxybenzoate in 36 mL 95% ethanol, 1.3 g streptomycin in 32.4 mL water, 15.2 mL propionic acid.
7. Aliquot beakers a–d into small Petri dishes and add vector to 1.5 L small vials.
8. Because the foods resemble one another and standard sucrose fly food, the vials, bottles, and dishes were carefully labeled with an easily remembered marker stripe system to distinguish each type.

METHODS

Generation of Isogenic Drosophila Strains

As previously discussed, it is important to use isogenized *Drosophila* strains for nutrigenomics studies. If one were to use the Exelixis PBac insertions or deficiencies, one could use them directly because they are already in the isogenized strain, w^{1118}, *iso-2; iso-3*. However, if mutations already exist in your favorite gene (*yfg*$^-$), but in another genetic background, one should backcross the mutation for several generations (typically 10–20 generations) to an isogenic strain and balance the mutations to balancer chromosomes in the same isogenetic background. Fortunately, the w^{1118}, *iso-2; iso-3* isogenic balancer strains can be ordered from the Indiana stock collection.

The *Drosophila* strains can be isogenized by "*washing*" the *yfg*⁻ mutation for several generations in the female germline—recombination of homologous chromosomes occurs only in female flies. We discussed these techniques last year in this series, but the release of the Exelixis isogenic PBac insertion strains, previously discussed, greatly simplifies the isogenization protocol. Now, in order to get a mutation in your favorite gene (yfg^-) in the isogenic background, simply backcross the strain with your mutation to the nearest PBac-transposon insertion, preferably one that is recessive lethal. Because of the large number of PBac insertion mutations, the PBac is very likely to be within much less than 1 cM (1% recombination) from your favorite gene, if not in the gene itself. The PBac insertions are conveniently marked with the mini- w^+ transgene, so one can select for PBac(w^+)/yfg^- in every generation of a backcross between yfg^-/PBac(w^+) virgin females and PBac(w^+)/Balancer (*Bal*) males. Notice that PBac(w^+)/PBac(w^+) flies will not be present if the PBac insertion is a recessive lethal mutation, as we suggested, and yfg^-/yfg^- flies will not be present if *yfg* is an essential gene. If the PBac insertion is not a recessive lethal, one can still use this approach, but one must be able to distinguish PBac(w^+)/PBac(w^+) flies from PBac(w^+)/*yfm* flies, which is usually very easy to do based on the intensity of pigment in the eyes. One can also "wash" the chromsome if *yfg* is not an essential gene using the same logic and approach as previously mentioned.

Triglyceride, Total Protein, and Live Weight Determination

Virgin males and virgin females were randomly collected from the vials and bottles and prepared as described previously in the Materials section to control for density. Six replicates per line per sex, each containing a pool of 10 flies, were used for the experiment. Twenty-four hours after collection, flies were weighed and homogenized. The homogenization protocol is the same as in Clark and Keith. Briefly, adults were homogenized on ice using 25 μL of homogenization buffer. The homogenates were centrifuged in a microcentrifuge at 2000 rpm for 2 min. The lipid layer on the surface was resuspended with the supernatant and the homogenate were distributed in 0.5 mL tubes. Live weight, triglycerides, and total proteins were measured for each sample, following the protocols listed as follows:

1. Live weight was measured to 0.1 mg accuracy with an analytical balance.
2. Triglycerides were assayed spectrophotometrically using the *Vitros* DT60 II reader and *Vitros* TRIG DT slides. The *Vitros* TRIG DT slide is a dry, multilayered film in a plastic support. It contains all the reagents necessary to determine triglyceride levels in the homogenate. A 10-μl drop of homogenate was deposited on the slide. The sample spreads evenly and diffuses into the film layers. Triglycerides in the sample undergo a series of reactions in the slide to produce a colored compound. The intensity of the color is proportional to the amount of triglycerides in the sample.
3. Total proteins were assayed using modified Lowry protein assay reagent kit. Aliquots of 200 μL of homogenate were added to 1 mL of Modified Lowry Reagent. After a 10 min incubation period at room temperature, 100 μL of prepared 1X Folin-Ciocalteu Reagent was added. After a 30 min incubation period at room temperature, A_{750} was measured. Protein contents were calculated relative to protein standards run with each replicate group.

Statistical Analyses for Triglyceride, Total Protein and Live Weight

For all three assays, analysis of covariance (ANCOVA) was used to detect genetic phenotypic variation due to food source and gender, with body weight and total proteins as covariates. The significant results (i.e., $p < 0.05$) of these metabolic experiments were that triglycerides were significantly lowered in female flies and pooled male and female flies after 10 d on the beef diet ($p = 0.023\ 3$ and $p = 0.0075$, respectively). Also, the total protein significantly increased almost two-fold in male flies fed soy for 10 d ($p = 0.0132$) and significantly increased by more than 50% in pooled flies fed beef for 10 d ($p = 0.001$). None of the other conditions produced significant effects in triglyceride levels or total proteins, and none of the three diets significantly affected live weight. The latter result is significant because it shows that it is not simply the weight of the flies that is being affected.

Measuring the Time Between Egg Laying to Eclosion as Adults

In the adult longevity studies, every 2 d the adults were transferred to fresh food in a new cage and the number of dead males and females were counted. The old food with eggs was cut into pieces containing no more than 50 eggs and placed into empty 25 mL vials to determine the fecundity of the parents and to measure the amount of time it

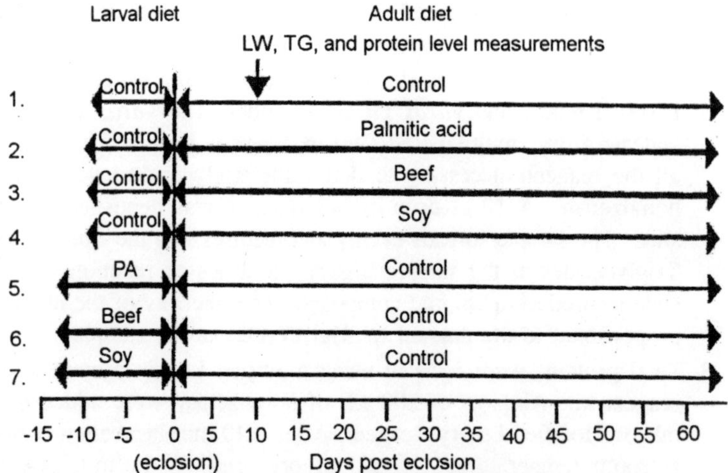

Fig. 5.1. Seven dietary conditions.

takes for the progeny to progress from egg laying to adult. Also, virgin males and females were collected from the vials at least twice a day and kept in separate vials with control food (no more than 20 male or female flies per vial) to determine the life span of these flies. We found that some dietary conditions during larval life affected the life span of the flies. However, these experiments are beyond the scope of this chapter and will be presented at a later time.

The time between egg laying to emergence was shortest for the control food (14.5 ± 1.5 d), and longest for the palmitic acid food (20.0 ± 2.8 d). Soy and beef diets caused intermediate developmental delays (16.8 ± 3.5 and 17.4 ± 2.3 d, respectively).

Longevity Studies With Mated Males and Females

About 100 freshly collected virgin flies (no more than 50 males and 50 females) with the isogenic background, w^{1118}, *iso-2; iso-3*, were raised in control food during their larval stages and transferred into a bottle with a food plate. Every other day, the flies were transferred to fresh food plates and the numbers of dead male and female flies were counted. The experiment was continued until all flies died. Survival analysis was performed using the Logrank and Wilcoxon tests calculated by SPSS.

At least three to five bottles of approx 100 flies were used for each of the five diets. In order to do this, approximately five bottles containing 50 mL of control food and approx 100 parental flies were set up about 2 wk before the flies were needed. The bottles were

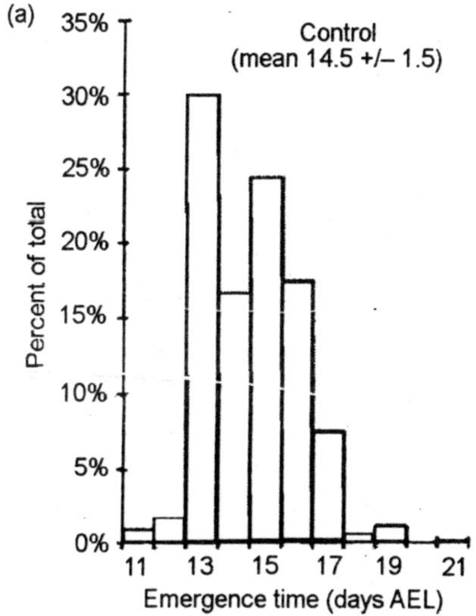

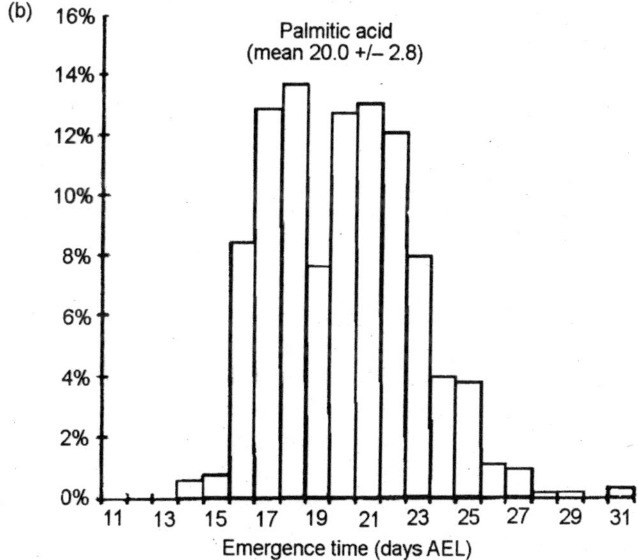

Fig. 5.2. The larval periods are extended in palmitic acid, soy and beef diets.

transferred every 2 d so that the larvae were about the same age. From this set of bottles, we could collect a total of between 100 and 300 virgin males and females every day, which were in turn used for the longevity experiments.

Note that this is a "*running*" experiment because we could not collect all of the virgin males and females needed for all of the experiments in 1 d. To keep track of the cages for the longevity experiment, we used paper with grid lines in a loose-leaf notebook, although an Excel spreadsheet could have been used if one prefers. The first cage was labeled "1.1 (date-1)", where "1.1" indicates "diet 1, cage 1," and "date" is the day that the cage was set up. The first page in the notebook had "1.1 (date-1)" labeling the first column, 1.2 (date-2) labeling the second column, and so on to the end of the page. The rows were labeled at 1 d intervals starting from the date that the first cage was set up, and 1-d intervals were listed down the rows to the bottom of the page. The second page of the notebook had "2.1 (date-2)" in a similar manner as above. We preferred using a binder rather than a notebook so that additional pages could be added, if necessary. Also, it is easier to work with paper and pen at the microscope than to use a computer in direct data entry.

The results are that mean and maximal longevity are significantly reduced for both males and females in all three diets compared with control diets, with the exception of females on the palmitic acid diet. The mean survival times, even in the control food, were considerably less than in previous studies. Probably the main reason for our lower observed survival times was that, because we wanted to analyze the offspring, the males and females were kept together and mated rather than virgin and separated, as is normally done. It is known that the longevity of reproducing male and female *Drosophila* is significantly decreased because of the "*cost of reproduction*".

It is also possible that life span is shortened in our experiments because we used large cages with approx 100 flies per cage vs other laboratories that use smaller numbers of flies in smaller vials. We found that when virgin males and females (the offspring of the flies used in these studies) are kept in separate vials (no more than 20 per 25 mL vial containing ~10 mL food), the life span is more than twice as long (data not shown). Nevertheless, we think that the relative longevity results are meaningful in the studies presented here because all of the flies were kept in identical environmental conditions, other than the diets, which was the variable under analysis.

While it is generally advised to use fly strains with long life spans for longevity experiments, this need not necessarily be the case. For example, Helfand's laboratory has recently shown that fly strains that have very short life spans because of genetic manipulations still have extended life spans under CR and other life span-promoting conditions. Therefore, even though, for practical reasons, the longevity experiments that we present in this chapter were not under optimal environmental or genetic conditions, we believe that the relative changes in life span under the different diets probably reflects what would happen if these experiments were conducted under other conditions.

Microarray Analyses

Each microarray analysis was done on 10-d-old adult flies, 40–50 flies (~50 mg) on each of the four diets for their entire adult lives. At least four microarray analyses were performed for each of the four diets (two for males and two for females). However, to increase the statistical power, the male and female data were combined for this chapter. Total RNA was extracted with TRIZOL and cleaned by RNeasy Mini Kit (QIAGEN). The SuperScript Indirect cDNA Labeling System (Invitrogen) was used to generate fluorescently labeled cDNA. The SuperScript System uses an aminoallyl-modified nucleotide and an aminohexyl-modifiled nucleotide together with other dNTPs in a cDNA synthesis reaction with SuperScript III RNase H Reverse Transcriptase. After a purification step to remove unincorporated nucleotides, the amino-modified cDNA was coupled with Cy3 or Cy5 dyes from Amersham Biosciences. Unreacted dye was removed and then labeled cDNA was hybridized to BDGP nearly-whole genome microarrays containing approx 13,000 probes. The microarray images were scanned, read and processed into data files by Genepix Pro software.

Data Preprocessing and Statistical Analysis of Microarray Experiments

As a result of the presence of multiple sources of bias, the raw microarray data were logarithm-transformed and normalized using "*joint lowess*" to remove both intensity and location biases from each array.

The overall mean of each channel on each array was subtracted from each data point to remove the overall difference between channels. The normalized data were then fitted to a simple fixed effect analysis of variance (ANOVA) model,

$$Y_{ijk} = \mu + D_i + S_j + A_k + \varepsilon_{ijk},$$

where μ represents the gene mean; D_i (I = 1,2) represents the diet effect, S_j (j = 1,2) represents the sex effect, A_k (k = 1, 2,..., number of arrays) represents the array effect; and the ε_{ijk} is the residual. The diet effect is tested using a shrinkage based t statistic. Because of the uncertainty of distribution of the residual error, ε_{ijk}, permutation analysis was used to establish the null distribution of the t statistic and the t statistics from 500 permutations were pooled across genes to provide a smooth distribution for obtaining *p* values. Because of the large number of genes tested in the experiment, the multiple testing adjustment procedure, *false discovery rate* (FDR), was applied to the *p* values. For all diets, FDR of 0.1, unless otherwise specified, were used to select significant genes.

Graphic Displays of Differentially Expressed Genes

Volcano plots for visualizing test results

The statistical test results for detecting the dye effect are visualized using volcano plots, which plot the fold change on the *x* axis and the statistical strength ($-\log_{10}$ of the tabulated *p* values of the conventional t test) on the *y* axis for each gene. The significant genes selected are highlighted (gray) points at the upper corners. A horizontal line is used to represent the nominal significance level of 0.001 for the conventional t test. By looking at a volcano plot, the fold change and the statistical significant of each gene and roughly how many genes are significant is evident.

Histograms of p values

The distribution of *p* values obtained from testing each gene provides information about whether there are differentially expressed genes in the whole experiment. If none of the genes are differentially expressed, the *p* values will have a uniform distribution. The excess of small *p* values indicates the presence of differentially expressed gene. Unlike a volcano plot, a histogram of *p* values can indicate some problems in the data. For example, the presence of multiple peaks may indicate the violation of test assumptions. Therefore, we often plot the histogram of *p* values for visual inspection.

Overlapping Differential Expression

Because of the limited amount of fly material in this pilot study, we did not have sufficient RNA from each sample to perform a dye-swap experiment as is normally done in two-channel array experiments. Thus, it is impossible for us to distinguish between effects due to diet vs effects due to labeling with different dyes. Nevertheless, we do not think that there are significant effects due to differential labeling with different dyes because, in contrast to direct labeling of probes with Cy3 and Cy5-labeled nucleotides, indirect labeling of probes is more efficient and, in our experience, does not introduce as much bias. With the above caveats, we note a number of potentially revealing

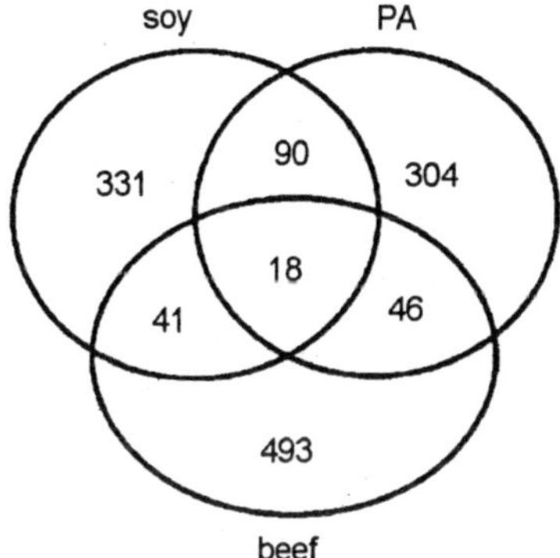

Fig. 5.3. Venn diagram of genes with altered expression in the three diets compared with the control diet (FDR = 0.1).

gene expression differences. Using an FDR cutoff of 0.1 for all three data sets and an additional *p* value cutoff of 0.003 for the beef data set (to reduce the number of candidate genes), we compiled a list of array elements and associated genes that appeared to be differentially expressed relative to the controls. We used the corresponding Flybase 'gn' codes (when available) to retrieve Gene Ontology (GO) functional annotations and identify candidate genes that may warrant future study. FlyBase ids were unavailable for a few (around 30 per experiment) array elements; we discarded these in the subsequent analysis because it is unclear what these array elements represent.

We examined the overlap between sets of differentially expressed genes between and among the different dietary data sets. The Venn tallies the number of genes found to be differentially expressed under each diet and combinations of diets. The intersection of all three diets contains 18 genes, including one false-positive consisting of several LacZ array element controls that were grouped under one Flybase "gn" id and which ought not to give a differential expression result. *Drosophila* do not have a *lacZ* gene, but their gut bacterial florae presumably do (based on positive X-gal staining), so it is possible that the bacteria grow to different levels when their hosts are in the different diets. We mention this clear-cut false-positive in order to emphasize that experiments such as these that involve many thousands of statistical tests almost certainly produce a number of false-positives that must be screened in subsequent experiments.

Until further work is done, we hesitate to draw conclusions regarding how the different diets affect global gene expression patterns; however, these results do suggest that DNA microarrays have the potential to implicate specific pathways as being particularly responsive to diet.

Future Prospects

The results of these studies are, surprisingly, that triglyceride and total protein levels are significantly decreased by the beef diet in all adults, and total protein levels are significantly increased in male flies fed the soy diet. Furthermore, we found that all three experimental diets significantly decrease longevity, increase the length of time to develop from egg to adult, and alter global gene expression patterns compared with the control high-sucrose diet. Preliminary microarray analyses suggest that a total of 60 genes have significantly (FDR = 0.1) altered gene expression diets by all three experimental diets compared with the control diet. The life-shortening effect of palmatic acid confirms previous studies by Driver's laboratory.

The triglyceride-lowering effects of beef are consistent with recent studies in humans comparing high-fiber (HC), high-fat (HF - Atkins), and high-protein (HP-Zone) diets. According to the authors, "Body weight, waist circumference, triglycerides and insulin levels decreased with all three diets but, apart from insulin, the reductions were significantly greater in the HF and HP groups than in the HC group". Soy isoflavones have also been reported to decrease triglyceride levels in mammalian models, but we did not observe this effect in *Drosophila* fed the high-soy diet.

We interpret that time required from egg laying to eclosion as being inversely proportional to the "*quality*" of food, at least as it applies to proper *Drosophila* larval development. If this interpretation is correct, then it suggests that the control (high sucrose) food also has the highest "*quality*." This is not surprising because fruit flies typically lay eggs in rotting fruit, hence their name. Rotting fruits are high in fructose and fermentation products, which like sucrose, are quickly broken down and easily digested. Likewise, we interpret the longevity results as indicating that sucrose has the highest "quality" in terms of longevity.

It might be that the other diets require more energy for digestion and energy expenditure might decrease life span. We do not think that this is the case because mouse models heterozygous for the insulin receptor have an increase in energy expenditure, but also an increase in mean and maximal life span. Therefore, "quality" likely means micronutrients that are essential for optimal *Drosophila* development. The sucrose-yeast diet is the highest "quality" possibly because, as mentioned previously, this diet most closely resembles the natural diet of *Drosophila*, to which they are adapted. One could test this idea by performing longevity and microarray analyses on insects that specifically eat soy beans, for instance, or insects that primarily eat meat, such as carrion beetles. The prediction would be that the shortest larval stages and the longest life span would be observed when these insects are reared on diets that most closely resemble their natural diets.

Another possibility is that components in soy, such as the isoflavones genistein and daidzein, interfere with ecdysone steroid-hormone signaling in *Drosophila*. Sharpe's laboratory, which has studied the effects of environmental estrogens on mammalian hormonal signaling for several decades, published an influential paper in 2002 that showed that soy formula decreases testosterone levels in male marmoset monkeys. Sharpe also suggested in a review on the "*estrogen hypothesis*" that this could lead to testicular abnormalities and cancer. The marmoset result was a primary reason for the recent recommendation to discourage the use of soy-based infant formula for babies under 6 mo of age in the United Kingdom.

Drosophila do not have estrogens or testosterones, but the steroid hormone ecdysone is the primary molting hormone that is required for inducing the larval molts and pupation. If soy isoflavones, for instance, interfere with ecdysone signaling, this could help explain why the times from egg laying to eclosion are prolonged when larvae are on the soy diet. However, this would not explain why similar prolongation effects are seen in the beef and palmitic acid diets. Also, a decrease in ecdysone signaling in adult flies *increases* longevity, whereas we see a *decrease* in longevity when flies are fed soy, which is opposite to the predicted result.

The preliminary microarray results are informative in determining how gene expression changes when *Drosophila* are fed the various diets. Many of the lipid catabolism genes have altered regulation in palmitic acid, beef, and soy diets. This is not surprising because all three of these foods are high in lipids. However, most of the genes that have altered gene expression profiles in the various diets are not as easily interpreted. For example, one intriguing finding is that several olfactory pathway genes have altered expression patterns in the various diets. It would be interesting to conduct behavioral experiments to determine if mutations in these genes affect preference for the various diets. Because of space limitations,we did not exhaust the types of microarray analyses that we could have performed. For example, the

Badger laboratory recently performed microarray experiments to analyze gene expression changes in particular tissues caused by rats fed a high-soy diet. A meta-analysis comparing gene expression changes in rats and *Drosophila* fed high-soy diets would be interesting. Also, the Partridge laboratory has published a comprehensive study on microarray analyses of aging and calorically restricted *Drosophila*. In this thorough and well conducted paper, they used several analysis techniques that we did not present in this chapter. For example, in addition to survivorship curves, these investigators also showed curves representing the mortality rates vs age. Also, they performed a time course of gene expression changes in control or calorically restricted conditions. Additionally, the analyses of the data in terms of gene ontogeny (GO) codes and the number of up- or downregulated genes with a particular GO code was more extensively determined in their paper than in this chapter.

The studies presented in this chapter also set the stage for studying the possible long-term "*metabolic imprinting effects*" of various diets. "*Metabolic imprinting*" is the hypothesis that long-term global alterations in chromatin organization, and therefore in the expression of genes, are induced by dietary components. During metabolic imprinting, it has been hypothesized that chromatin alterations persist long after the inducing components are removed from the diet, thereby causing a stably altered metabolic state. Some evidence suggests that metabolic imprinting can persist for a significant portion of an organism's life span, and possibly even in subsequent generations,by heritable-epigenetic alteration of genes. It has been proposed that some influences on obesity may occur *in utero* by the mother's diet or even by the grandmother's diet. Further evidence that trans-generational effects of obesity might be regulated by epigenetic (e.g., DNA methylation) events is the recent finding that cloned mice tend to be obese yet do not pass on this obesity to their offspring.

We note that the interpretation of the diet experiments presented in this chapter is arguably limited by our choice of using only one particular concentration of each nutrient (sucrose, palmitic acid, soy, or beef). Life span data from flies maintained on a range of concentrations of each nutrient might be required before a meaningful inference can be developed. For example, one can examine published data from flies experiencing dietary restriction. When high-nutrient medium composed of 15% w/v yeast-sucrose medium is diluted, life span increases while reproduction decreases (the dietary restriction effect) until the food level reaches roughly 5–7% w/v concentration. Further dilution results in a decrease in life span, presumably as a result of malnutrition. Thus, there is a parabolic (concave down) relationship between sucrose-yeast concentration and life span. Therefore, one could logically conclude that two diets could be chosen such that the lower concentration is either short lived, long-lived, or has identical lifespan in comparison to higher concentration.

However, an argument supporting our approach of using different isocaloric diets at only a high concentration is that it has recently been reported that dilution of food is, in many cases, almost fully compensated for by an increase in food uptake. Carvalho and colleagues demonstrated that dietary restriction elicits robust compensatory changes in food consumption. Therefore, feeding behavior and nutritional composition act concertedly to determine fly life span. We argue that feeding isocaloric diets of different compositions, as we describe in this chapter, might be a better controlled experiment than dietary restriction by dilution favored by the majority of the longevity scientific community. Although both dietary restriction and dietary composition studies assume equal volumes of food intake, evidently an incorrect assumption in the dietary restriction studies, it is possible that this assumption is valid with the isocaloric diets described in this chapter. We advocate that whatever approach is used, proper studies are needed to control for the volume of food intake.

Despite the obvious benefits illustrated previously of using *Drosophila* as a model to combine nutrigenomic and longevity studies, one might still have valid arguments against conducting these studies in *Drosophila*. One could raise the point, for instance, that we use in our studies a variety of "*nonnatural*" food sources (palmitic acid, soy, beef), and we find that they slow down development and decrease adult lifespan. Because *Drosophila* eat yeast, this is unsurprising because they presumably suffer some degree of malnutrition on these unnatural food sources. However, this argument misses the main point of these studies, which is ultimately a comparative metabolic study between humans and *Drosophila*. *Drosophila* in the wild eat primarily yeast on fermenting fruit, and tofu and beef are not "*natural*" foods for this species, but what is a "*natural*" food for humans? Elaine Morgan, in her controversial book *The Aquatic Ape: A Theory of Human Evolution*, argues that "hairless" human ancestors, like dolphins and whales, were originally aquatic animals who ate primarily fish. Evidence cited for this controversial hypothesis is that humans must have enough eicosanoic acid (a 20 carbon mono-unsaturated fatty acid) and other omega-3 fatty acids, primarily found in fish, for proper brain and body development. Regardless of which of the many hypotheses for human evolution are correct, tofu and beef are almost certainly "nonnatural" diets for both humans and *Drosophila*. The unnaturalness of many human foods, and especially extreme diets favored by many in the United States, further illustrates the need to conduct comparative nutrigenomics studies with both "*natural*" and "*nonnatural*" foods.

Nevertheless, one should be careful in applying what has been learned about *Drosophila* nutrigenomics to humans. Dietary requirements are certainly quite different between flies and man, and even between two insect or mammalian species. Nevertheless, we believe that our findings contribute to the field of nutrigenomics, and will help identify evolutionarily conserved mechanisms connecting diet to metabolism.

6

DNA METHYLTRANSFERASE IN AGING

DNA methylation of the 5-position of cytosine is considered to be the most common modification of the genome in mammals, and methylation in regulatory regions of genes has been shown to be inversely related to gene expression. Methylation patterns of cytosines are associated with many cellular processes, including chromatin structuring, development, carcinogenesis, and aging. DNA methyltransferase (Dnmt) catalyzes the transfer of a methyl moiety from *S*-adenoslyl-L-methionine (SAM) to the 5-postion of cytosines in the CpG dinucleotide. Several Dnmts have been identified in somatic tissues of vertebrates. Dnmt1 is the most abundant methyltransferase in mammalian cells and is primarily associated with maintaining established methylation patterns in newly synthesized DNA.

Methylation patterns have been shown to be important in normal embryonic development. For example, monoclonal antibodies against Dnmt1 halt cell division in early-stage *Xenopus* embryos. Morphological data suggested that cells were arrested in interphase, and this was supported by biochemical analysis that indicated cells had not entered M-phase. Dnmt1$^{-/-}$ and Dnmt3b$^{-/-}$ mutations in mice result in embryonic death while mice with mutations Dnmt1,3a$^{-/-}$ die at approx 4 wk of age. A study investigating DNMT activity in human systems showed that overall maintenance methylation decreased during cellular senescence of WI-38 fibroblasts and that combined *de novo* methylation initially decreased but later increased as these cells aged. Thus many recent studies have illustrated the important role of Dnmts in embryonic development and aging. However, much remains to be elucidated and research aimed at Dnmt activities both in vivo and in vitro is essential to understanding the mechanisms of important biological processes such as aging.

MATERIALS

RNA Extraction and Quantification

Items 1-4 are included in the Qiagen RNeasy Mini Kit.

1. RNeasy mini spin column.
2. Buffer RLT.
3. Buffer RPE.
4. Buffer RW1.
5. QIAshredder Qiagen (cat. no. 79656).
6. β-mercaptoethanol.
7. Phosphate-buffered saline (PBS).
8. Ethanol (both 70% and 100%).
9. Spectrophotometer Cuvet Silica (quartz).
10. SmartSpec Plus spectrophotometer.

Synthesis of cDNA and Sequence-Specific PCR

Items 1-4 are included in the SuperScript First-Strand Synthesis System for reverse-transcription (RT)-PCR kit.

1. Control RNA.
2. 50 ng/μL random Hexamer mix.
3. 10 m*M*dNTP mix.
4. Diethyl pyrocarbonate (DEPC)-treated water.
5. Thin-walled 0.2 mL PCR tubes.
6. Gene-specific primers.
7. PCR Master Mix.
8. Nuclease-free water.

Gel Electrophoresis

1. Gel electrophoresis apparatus.
2. Gel electrophoresis-grade agarose.
3. 1X TAE buffer.
4. 100 bp DNA molecular marker.
5. 6X Loading Dye.
6. 10 mg/mL ethidium bromide solution.
7. Ultraviolet (UV) trans-illuminator.

Methods

Cell Extract Preparation

The following protocol pertains to adherent aging cell cultures, for cells grown in suspension.

1. Aspirate media from experimental and control cells.
2. Wash bottom of flasks with a predetermined amount of 1X PBS warmed to 37°C.
3. Add a predetermined amount of trypsin-EDTA directly to the bottom of the flasks and incubate at 37°C for 5–6 min.
4. Collect detached cells by washing the bottom of the flasks with media warmed to 37°C and add the media/trypsin/cell mix to a 15-mL conical tube for centrifugation.
5. Pellet cells by centrifugation and resuspend pellets in PBS. Count cells using a method determined by the experimenter.
6. Place 1-3 × 10^6 cells in a 1.5-mL tube and pellet by centrifugation. Remove all PBS from tubes. Pellets may be frozen at –80°C for later use.

Cell Lysis and Homogenization

1. Prepare cell lysis buffer solution by adding 10 μL of β-mercaptoethanol to 1 mL of buffer RLT.
2. Disrupt pellet obtained as described above by flicking the tube and pipet 350 μL of freshly prepared cell lysis buffer solution to cell pellets obtained as described anpve. It is important to achieve a substantial amount of disruption to the pellet in order to accomplish complete lysis.
3. To homogenize, pipet lysate into a QIAshredder placed in a 2-mL collection tube and spin for 2 min at maximum speed in a microcentrifuge.

RNA Extraction

Because of the instability of RNA, it is necessary to perform the following protocol quickly and without too much pause between steps. It is also not recommended to extract RNA from more than six to seven samples at one time.

1. Add 350 μL of 70% ethanol to homogenized lysate and mix thoroughly by pipetting. Pipet the mixture up and down at least 10 times to mix. Apply the entire sample (700 μL) to an RNeasy mini spin column and centrifuge at 8000*g* for 30 s.

Table 6.1. RNA and primer mix

Reagent	*Sample reaction*	*Positive RT control*
1-3 μg Sample RNA	*n* μL	—
Control RNA	—	1 μL
Random hexamers	1.5 μL	1.5 μL
dNTP mix	1.5 μL	1.5 μL
Water	to 10 μL	to 10 μL

2. Discard the liquid in the collection tube only and add 700 μL of buffer RW1 to the column placed in the same collection tube and centrifuge at 8000*g* for 30 s.
3. Transfer the RNeasy column to a new collection tube and discard the previously used tube containing flow through from step 3.
4. Prepare buffer RPE by adding 4 volumes of 100% ethanol to 1 volume of RPE concentrate. Add 500 μL of the prepared RPE buffer to the RNeasy column and centrifuge at 8000*g* for 30 s.
5. Discard the liquid in the collection tube and pipet another 500 μL of the prepared RPE into the RNeasy column. Centrifuge at 8000*g* for 2 min.
6. Place RNeasy column in a new collection tube and centrifuge for 1 min at maximum speed.
7. Place RNeasy column in a 1.5-mL collection tube and elute RNA from column by adding 30 μL of RNase-free water. Let column stand for 1 min and centrifuge at 8000*g* for 1 min.

RNA Quantification

1. Be sure to clean the silica cuvet with 1 wash of 70% ethanol followed by two to three washes of sterile water.
2. Add 495 μL of sterile water to the cuvet. Blank the spectrophotometer. Pipet 5 μl of extracted RNA obtained. Mix by gentle inversion.
3. Repeat step 2 for each sample to be quantified.
4. Spectrophotometer concentration reading is micrograms per milliliter. Convert reading to micrograms per microliter.

First-Strand cDNA Synthesis

1. For each RNA sample obtained, prepare a mixture containing reagents in Table 6.1 (*n* is the amount of sample required to achieve desired concentration of RNA needed for conversion to cDNA).

2. Incubate each sample at 65°C for 5 min followed by short incubation on ice.
3. During the incubation in step 2, prepare enough reaction master mix for all samples, including control.
4. Pipet 9 μL of the reaction master mix into each tube containing mixture from step 1, mix briefly by vortexing, and collect samples by brief centrifugation.

Table 6.2. cDNA reaction master mix

Reagent	*For each reaction*
10X reverse-transcription buffer	2 μL
25 m*M* $MgCL_2$	4 μL
0.1 *M* dithiothreitol	2 μL
RNaseOUT Ribonuclease Inhibitor	1 μL

5. Incubate at 25°C for 2 min.
6. Add 1 μL of SuperScript II reverse transcriptase to each tube and mix gently and collect by brief centrifugation.
7. Incubate at 25°C for 10 min.
8. Transfer tubes to 42°C for 50 min.
9. To terminate the reactions, incubate at 70°C for 15 min and chill on ice.
10. Collect the samples by brief centrifugation and add 1 μL of RNase H to each tube. Incubate at 37°C for 20 min. Samples are now converted to cDNA and are ready for sequence specific PCR.

Table 6.3. Sequence-specific PCR

Reagent	*For each reaction*
2X PCR master mix	15 μL
Nuclease-free water	9 μL
Forward: 5'-ACCGCTTCTACTTCCTCGAGGCCTA-3'	3 μL of forward and
Reverse: 5'-GTTGCAGTCCTCTGTGAACACTGTGG-3'	reverse primer mix

DNMT 1 Gene Amplification by PCR

1. Prepare a master mix containing the reagents listed in **Table 3**.
2. For each reaction, add 3 μL of cDNA sample obtained as described above to a 0.2-mL PCR tube containing 27 μL of prepared master mix.
3. Mix gently and collect by brief centrifugation.
4. Perform PCR as follows: 94°C for 5 min; 33 cycles of 94°C for 30s, 56°C for 40 s, 72°C for 40 s followed by 72°C hold for 7 min.

Gel Electrophoresis

1. Prepare a 2% agarose gel by mixing 2 g of agarose with 100 mL of TAE buffer.
2. Load 2 μL of 6X loading dye and 10 μL of sample obtained as described above, into each well.
3. Run gel for approx 30 min at 100 V. Alternatively, gels can be run at 50 V for approx 1 h.
4. Mix 20 μL of ethidium bromide with approx 100 mL of TAE buffer and incubate gel for 15 min. Make sure to mix the ethidium bromide/TAE solution thoroughly to ensure even distribution before incubating gel.

Notes

1. Free-floating or suspension cells require a slightly different technique for extraction. It is not necessary to detach the cells using trypsin. Cells must be collected and placed in a 15-mL conical tube and then pelleted by centrifugation.
2. The amount of trypsin added to the plates is determined by the size of the dish used in the experiment. For example, to achieve complete detachment of cells from a 75-mm^2 dish, add 3.5 mL of trypsin followed by incubation at 37°C for 5-6 min. It has been shown that weekly trypsinizations do not affect profliferative capacity of the cell culture; however, because trypsin is a harsh digestive enzyme, the cells should not be left exposed for longer than necessary. Trypsin is inactivated by the serum component in media and detached cells should be suspended in 5-6 mL of media. The amount of PBS used also varies according to dish size. Use enough to cover the bottom of the dish to effectively wash away media.
3. Cell counting can be achieved by several different methods. Counting using a hemacytometer is a simple way to determine an approximate cell count. Cells are suspended in PBS as described above and vortexed for approx 1 min to disassociate clumps. Twenty-five microliters of cell suspension is added to a 1.5-mL collection tube and mixed with 75 μL of Trypan Blue dye. Nine microliters of the cell/trypan blue mix is added to the hemacytometer. Cells that are stained dark blue are considered not viable and should not be counted. The use of flow cytometry is also a method of cell counting. Volumetric flow cytometry is used to count absolute cell number in a given volume and provides a more accurate count than counting cells with a hemacytometer.

Expression of DNA Methyltransferase in Aging

Immunoblot analysis is a common tool used to provide a semiquantitative measurement of both protein expression and stability. Proteins have been known to undergo many posttranslational modifications which include acetylation, ubiquitination and phosphorylation, and indeed many studies have used this technique to study these protein alterations. DNA methyltransferases (DNMTs) are important enzymes involved in transferring methyl moieties to the 5-postion of cytosine in CpG dinucleotides and thereby regulating gene activity. In addition to changes in transcription of DNMTs, protein levels have also been shown to fluctuate as cells age and enter senescence. Moreover, it has also been revealed that DNMTs can undergo posttranslational modifications, which indicates the importance of studying DNMTs at the protein level. The following describes a technique used in our laboratory to study DNMT protein expression in aging systems. In brief, total cellular protein is isolated and quantified, followed by denaturation and sodium dodecyl sulfate (SDS)-polyacrylamide gel electrophoresis (PAGE). The separated protein fraction is then transferred to a membrane and incubated with antibodies specific for the protein or proteins of interest. The probed membrane is subjected to a technique for visualization.

Table 6.4. Suggested protease inhibitors for use during sample preparation procedure

Inhibitor	*Working concentrations*	*Inhibition*
Benzamidine	1 m*M*	Trypsin-like serine proteases (thrombin, plasmin)
Leupeptin	1–10 μ*M*	Serine protease, plamsin and cystine proteases
Aprotinin	10 mg/mL	Serine protease, and plasmin
Phenylmethylsulfonyl flouride (PMSF)	1–2 m*M*	Serine protease with a trypsin-like specificity

Materials

Sample preparation

1. Phosphate-buffered saline.
2. Trypsin-EDTA: 0.25% w/v trypsin/0.2% w/v EDTA in sterile Hank's balanced salt solution.
3. Protease inhibitors.
4. NE-PER kit.
5. Bio-Rad Bradford Protein Assay.

SDS-polyacrylamide gel electrophoresis

1. 2X SDS loading buffer: 250 m*M* Tris pH 6.8, 4% SDS, 10% glycerol, 0.006% bromophenol blue, 2% β-mercaptoethanol.
2. Prestained protein molecular weight marker.
3. 5% precast SDS polyacrylamide gel.
4. 1X Running Buffer: 3.03 g Tris, 14.41 g Glycine, 1.5 g SDS in 1 L water.

Transfer and western blot

1. 1X transfer buffer: 3.03 g Tris, 14.41 g Glycine, 100 mL methanol in 1 L water.
2. Extra thick precut Whatman filter paper.
3. Sponge pads.
4. Hybond-ECL nitrocellulose membrane.
5. Heat-sealable plastic bags.
6. Nonfat dehydrated milk.
7. TBST: 10 m*M*Tris-HCl, pH 8.0, 150 mM NaCl and 0.1% Triton X-100.
8. Primary antibodies.
9. Horseradish peroxidase (HRP)-conjugated anti-immunoglobulin (Ig)G secondary antibodies.

Table 6.5. Primary antibodies used for expression analysis of the primary DNA methyltransferases (DNMTs)

DNA methyltransferase	*Target protein size*	*Specificity*
DNMT1	170–195 kDa	Goat polyclonal IgG
DNMT 3a	101–120 kDa	Goat polyclonal IgG
DNMT 3b	98–110 kDa	Goat polyclonal IgG

Detection

1. ECL kit.
2. Hyperfilm ECL.

Methods

Sample preparation

The majority of aging cells are grown in monolayer and can easily be harvested for DNMT protein level analysis with trypsin digest. The use of protease inhibitors during the process of harvesting cells and extracting protein is customary. While there are many methods to extract cellular protein (i.e., RIPA), the NE-PER kit allows for rapid isolation of nuclear and cytoplasmic extracts and is recommended in this case to extract nuclear DNMTs.

1. Grow cells to be examined to approx 70–80% confluence. Wash cells with cold PBS. Harvest monolayer cells with trypsin-EDTA and pellet cells at 300-350 × g (1500 rpm in a swinging bucket or 45° fixed-angle rotor) for 5 min at 4°C.
2. Wash pellet with cold PBS containing protease inhibitors.
3. Lyse cells and extract protein according to NE-PER kit protocol.
4. Protein can be flash-frozen in liquid nitrogen for future use.
5. Perform protein determination with the Bradford Assay.

SDS-polyacrylamide gel electrophoresis

Because DNMTs range in molecular weight from 98 to 195 kDa, there is no need to separate these peptides using the Tris-Tricine buffer system. The following method describes the traditional Laemmli discontinuous gel system which utilizes 0.1% SDS to denature the proteins. With pore size in the polyacrylamide gel matrix decreasing with higher acrylamide concentrations, the optimal gel concentration for DNMTs to migrate near the midpoint of the gel is 5%.

1. Boil 100 μg of protein in an equal amount (1:1 v/v dilution) of 2X SDS loading buffer for 10 min.
2. Load prestained marker and samples into the wells of a 5% SDS polyacrylamide gel.
3. Electrophorese 0.75-mm thick gel at 135 V of constant current until the bromophenol blue tracking dye reaches the bottom of the separating gel.

Transfer and western blot

DNMTs can be electrophoretically transferred in a tank for wet transfer or semi-dry apparatus to a nitrocellulose, polyvinyl difluoride (PVDF), or nylon membrane. Hybond-ECL membranes perform well for this assay. Incubations are best performed with low volumes (2-5 mL) in heat-sealable plastic bags and on orbital shakers. If plastic trays are used, increase volumes to 20-50 mL to ensure that membranes are submerged and always handle membrane with clean, gloved hands.

1. Set up transfer apparatus, taking care to equilibrate gel, transfer components, and activated membrane in transfer buffer at room temperature for 30 min and rolling out any air bubbles between filter paper, gel, and membrane with a glass rod or test tube.
2. Transfer at 100 V for 30–80 min at 4°C.

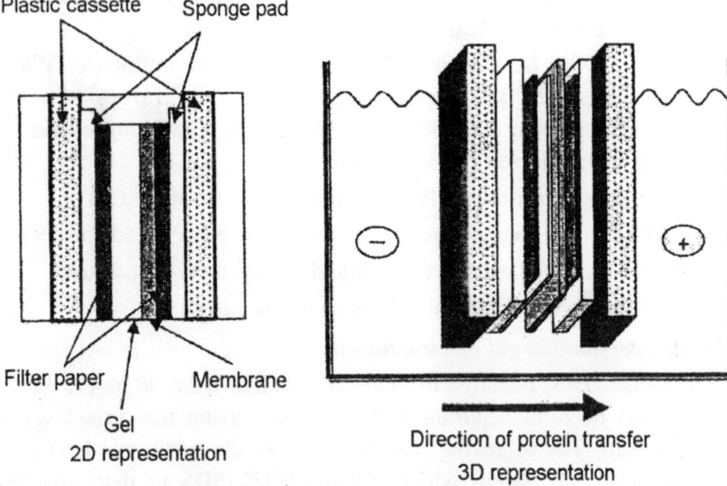

Fig. 6.1. Immunoblot transfer apparatus.

3. Turn off power supply, disassemble transfer apparatus, remove membrane and block in a solution of TBST for 30 min to 1 h at room temperature with gentle agitation or 4°C overnight. Alternatively, membrane can be blocked in 5% milk or 5% milk plus .01 mg/mL bovine serum albumin (BSA) for 2 h at room temperature or overnight at 4°C.
4. Dilute primary antibody in blocking buffer (determine dilution empirically). Probe membrane with primary antibody for at least 1 h at room temperature or overnight at 4°C. Place membrane in plastic tray and wash (with agitation) three times with TBST, 15 min per wash. Dilute secondary antibody in blocking buffer. Probe membrane with diluted secondary antibody in a new heat-sealable bag for 30 min to 1 h at room temperature with agitation.
5. Wash membrane 3X with TBST as in step 4.

Detection

Bound antigens can be visualized with chromogenic (i.e., BCIP/NBT) or chemiluminescent substrates. It is preferable to use HRP-conjugated secondary antibodies for the increased sensitivity of enhanced chemiluminescence. The Amersham ECL (Enhanced Chemiluminescence) system works well for this assay. Steps outlined in this section should be performed in a darkroom.

1. Remove traces of final wash by holding membrane vertically and dabbing edge of membrane to Whatman/absorbent paper.

2. Mix two substrate components (1:1 ratio) in a small plastic tray and lay membrane target-side down in this mixture for 1 min.
3. Place membrane target-side down in the center of a piece of plastic wrap and fold plastic wrap around membrane to create a flat, liquid tight seal.
4. Expose membrane to autoradiogram (X-ray) film for the appropriate time (5 s to 30 min, depending on the signal strength).
5. Develop autoradiogram, rinse membrane in PBS and air-dry.

Notes

1. If interested in two or more proteins with different molecular weights, use of a gradient gel is more efficient. Run gels at 150 V for 30 min or more depending on power pack. If gels are run at 200 V it does not yield a gel suitable for immunoblotting. In general, gel concentrations for proteins 60-600 kDa should be approx 5%, 16-70 kDa approx 10% and 12-45 kDa approx 15%.
2. Equilibrate the gel and nitrocellulose in transfer buffer that has 10% methanol for 15-30 min at room temperature. This allows the gel time to swell before it is transferred to a membrane. If this step is not done, the gel will swell during transfer and protein will not properly transfer to the membrane.
3. It is highly recommended that transfer occur for less than 80 min. Longer exposure to transfer buffer my lead to increased amounts of protein leaching out of the gel. Also, if using two complete transfer apparatuses, make sure that the power pack is able to produce the correct amount of voltage. Some of the older power supplies can not handle two setups and will take longer to complete the transfer. It has been noted that such immunoblots are usually of poor quality.
4. Primary antibodies are either polyclonal, monoclonal, ascites, or antiserum bleeds. Monoclonal antibodies are usually used at a starting dilution of 1:1000 but can go up to 1:10,000 or more. Polyclonal antibodies can range from a dilution as concentrated as 1:1 to a dilution of 1:500. Ascites antibodies are more concentrated than the monoclonal, so suggested starting concentration is 1:2000. Antiserum is rarely used for immunoblotting unless it has been purified by peptide/column or nitrocellulose. It generally gives high background and should only be used if other options are not readily accessible. It is recommended that different dilutions should be tested in the beginning for all antibodies to find the exact dilution needed.

5. Secondary antibodies are usually commercially made and come with a suggested dilution. It is advised to first try what the company recommends before changing dilution. Secondary antibodies should be chosen based on the primary antibody used. If probing the membrane for more than one primary antibody, note that the background will be extremely high. It is recommended that you probe for one antibody at a time or use a stripping agent after each detection.

7

DNA MICROARRAY TECHNOLOGY

The emergence of microarray technology as a novel tool in molecular biology has led to significant progress in many biomedical disciplines. Although no established classification exists for microarray subtypes, it is generally accepted that DNA microarrays consist of two categories. cDNA microarrays (printed on glass slides or on nylon filter membranes) rely on collections of bacterial clones containing inserts typically 500–2000 nucleotides long, whereas oligonucleotide microarrays rely on synthesized molecules 20–25 to 70–80 nucleotides long printed over the solid surfaces. Both DNA array formats have numerous individual advantages and disadvantages, related to the expenses and convenience associated with the manufacturing of platforms, probe synthesis, and hybridization process. Oligonucleotide arrays rely on biophysically optimized sequences. Being spotted with constant concentration across the array membrane, oligonucleotide clones generally provide reproducible and accurate quantitative data. On the other hand, application of long cDNA sequences results in less cross-hybridization of unrelated sequences, and although cDNA microarray are considered more labor-intensive, the latter might be preferential for certain applications.

An emerging class of microarrays relying on extremely long oligonucleotides might become a format of choice in the following decade. Similarly, certain advantages and disadvantages are associated with both radioactive and fluorescent labeling techniques. Although highly effective, radioactive labeling is relatively inexpensive, and membranes could be stripped and reprobed, if necessary. However, the ability to perform simultaneous two-color hybridizations (e.g., for the experiment vs control) and critical safety issues make the fluorescent labeling approach attractive for many researchers. Importantly, a number of DNA microarray platforms are currently available commercially or via custom design, providing a diversity of choices.

Over the course of just a few years, numerous technological advances have led to the evolution of DNA microarrays, which previously contained hundreds of spots of DNA material and now contain tens of thousands of these, reaching a truly genomic scale. In a number of gerontological studies, DNA microarrays have proved themselves as a powerful, high-throughput gene expression analysis tool. One group of studies has focused on the identification of basic physiological mechanisms of aging, whereas others aimed to uncover molecular mechanisms underlying the biological effects of anti-aging drugs. Ultimately, application of DNA microarray would greatly improve our understanding of mechanisms of aging and could result in identification of novel prognostic markers and therapeutic targets. Although the design and manufacturing of DNA microarray platforms and principles of data analysis are outside of the scope of this article, the protocols below embrace the actual handling of DNA microarrays in great detail.

Materials

Nylon Membrane-Based cDNA Microarray Hybridization Protocol, Radioactive Labeling

1. MicroHyb solution.
2. Mouse Cotl DNA (Invitrogen); store at -20°C.
3. PolyA RNA reconstituted to 50 μg/μL in 10 m*M* Tris-HCl, pH 7.5; store aliquots at -20°C.
4. High purity Oligo(dT)$_{12\text{-}18}$ primer (500 ng/μL).
5. RNasin ribonuclease inhibitor (Promega).
6. SuperScript II reverse transcriptase, supplied with 0.1 *M* dithiothreitol (DTT) and 5X 1st Buffer.
7. dNTPs: 100 m*M* stocks, diluted to 0.5 *M* each in molecular grade dH_2O (1 μL each of 100 m*M* dATP, dGTP, and dTTP and 197 μL of dH_2O); store at -20°C.
8. [α^{33P}] dCTP 10 mCi/ml.
9. 2X SSC, 2X SSC + 0.1% sodium dodecyl sulfate (SDS), 1X SSC + 0.1% SDS and 0.8X SSC + 0.1% SDS solutions prepared

using 20X *saline-sodium citrate* (SSC) buffer and 20% SDS solution stocks.

10. MicroSpin G-25 or G-50 columns.
11. 1 *M* Tris-HCl (pH 7.5), 0.5 *M* ethylenediamine tetraacetic acid (EDTA), and 0.1 *M* NaOH solutions.
12. Phosphor screen.
13. Scintillation counter.
14. Geiger counter.

Oligonucleotide Microarray Hybridization Protocol, Fluorescent Labeling

1. Low RNA Input Fluorescent Linear Amplification Kit. This kit contains reagents necessary for the synthesis of fluorescent cRNA, including enzymes, buffers, etc. Store at –20°C.
2. Cyanine 3-CTP (Cy3) and Cyanine 5-CTP (Cy5) (Perkin-Elmer). Store at 4°C, and protect from light.
3. RNeasy Mini-Kit. This kit contains RNeasy mini columns and essential reagents.
4. Pronto! Universal Hybridization Kit. This kit contains reagents necessary for pre-hybridization, hybridization of microarrays and post-hybridization washes. Wash Solutions 1–3 should be prepared in advance as the following:
   - (a) Wash Solution 1: mix (in the order indicated) 1074 mL of dH_2O, 120 mL of Universal Wash Reagent A, and 6 mL of Universal Wash Reagent B, aliquot to 500-mL volumes.
   - (b) Wash Solution 2 (prepare two 2-L bottles): mix (at the order indicated) 1710 mL of dH_2O and 90 mL of Universal Wash Reagent A.
   - (c) Wash Solution 3 (prepare two 2-L bottles): mix (at the order indicated) 360 mL of Wash Solution 2 and 1440 mL of dH_2O.
5. Ultraviolet crosslinker (UVP).
6. Hybridization chambers (Corning); should be cleaned after each hybridization and washed with RNase AWAY (Invitrogen) and 70% ethanol prior to the hybridization.

Methods

As a result of the diversity of DNA microarray platforms available, no single established protocol or reagent could be considered "*universal*" for every DNA microarray application. Instead, variations and adaptations of basic protocols are common at every stage of a microarray

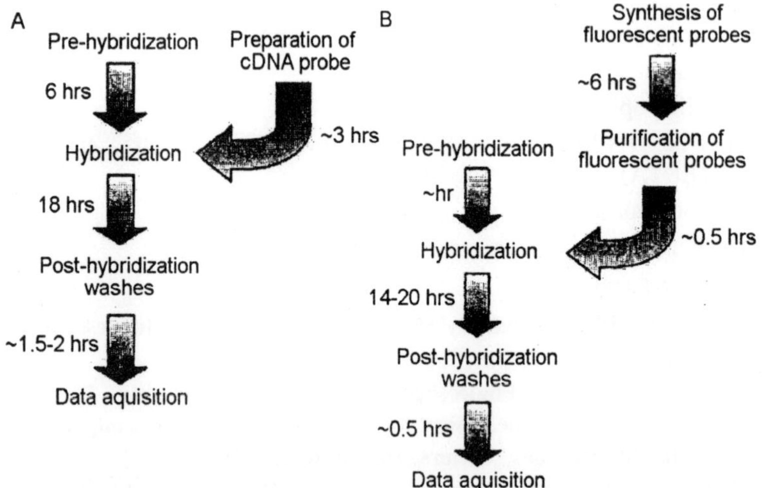

Fig. 7.1. Flowchart representing the sequence of major steps in the proposed methods. (A) cDNA microarray; radioactive labeling. (B) Oligonucleotide microarray; fluorescent labeling.

experiment, namely, sample labeling, pre-hybridization, hybridization, and post-hybridization washes. Below, details are given of two different protocols covering both cDNA and oligonucleotide microarray platforms, with radioactive and nonradioactive (fluorescent) labeling, providing a background for the technical aspects of everyday work with DNA microarrays.

RNA quality is critical for the success of microarray experiments and extensive effort should be made to validate RNA quality prior to launching the protocol. On all steps of RNA purification and handling, always use diethylpyrocarbonate (DEPC)-treated plasticware, aerosol barrier pipette tips and a dedicated set of pipets; frequently treat work surfaces and gloves with RNase AWAY; use rooms or chemical hoods dedicated to RNA work, when possible. Although RNA synthesis (including that of mitochondrial RNA) has long been known to decline with aging, no practical difficulties are associated with purifying the volume of total RNA required for microarray applications from the majority of samples. If necessary, micro-sample RNA isolation technique should be applied, followed by the RNA amplification.

Timing is particularly important in many steps of DNA microarray experiments. For example, if the pre-soak time is decreased below or extended beyond the 20-min period recommended for the fluorescently labeled oligonucleotide microarray hybridization protocol presented below, it could significantly reduce the resulting hybridization signal. This concern is most important when numerous microarray membranes or slides are handled simultaneously. It is therefore highly recommended that a researcher handle no more than one to three micro-arrays at once, until he or she is fully confident with a selected protocol.

Ultimately, no "*gold standard*" can be suggested for design of the experiments and statistical analysis of experimental results. Particulars should therefore be adjusted in each individual case according to the size of DNA microarray platform, the number of samples compared, etc. Moreover, technological means employed for data acquisition (i.e., scanner model, software available, etc.) and preferential mathematical means used for data analysis vary in different laboratories. The corresponding sections of protocols below are therefore provided in less detail and are for orientation purposes exclusively. Mathematical/statistical analysis of experimental data derived from microarray experiments is a subject of heated discussion, and numerous options for data analysis could be considered in each individual case. It had been found that cross-laboratory variations have a higher impact on the accuracy and precision of microarray data, compared to platform variations.

Standardized approaches of precision assessment and means of the statistical analysis are now under development, promising improved reproducibility of results. In conclusion, it should be stated that microarray hybridization results should always be validated using an alternative approach, such as conventional reverse-transcription (RT) PCR technology, real-time PCR (Q-PCR), or other. It is recommended that an assay of a limited (10–20) number of genes representing various levels of expression and/or expression alterations (i.e., upregulation vs downregulation in the experiment addressed) be studied. Trends detected should be compared to those observed in microarray analysis, with levels of agreement generally informing on the quality of microarray experiment.

Nylon Membrane-Based cDNA Microarray Hybridization Protocol, Radioactive Labeling

Pre-Hybridization

1. Pre-warm MicroHyb solution to 42°C, mixing intermittently. Consider it ready to be used when fully transparent (pre-warming might take 3–4 h).
2. Denature Cot1 DNA (1 mg/mL) and PolyA RNA (50 μg/μL) by boiling these for 2 min, and chill on ice for 2–3 min.

3. Rehydrate microarray membranes by placing them into a 50 mL volume of 2X SSC solution and incubating for 3 min at room temperature.
4. Mix 10 mL of pre-warmed MicroHyb solution with 250 μL of denatured Cot1 DNA (1 mg/mL) and 5 μL of denatured PolyA RNA (50 μg/μL).
5. Using forceps, place rehydrated membrane in the hybridization chamber, and add pre-hybridization mix.
6. Pre-hybridize membranes in a rolling hybridization chamber at 42°C for 6 h.

Preparation of the cDNA probe

1. Dilute or concentrate total RNA sample (20 μg) to 16 μL volume.
2. Add 2 μL of Oligo$(dT)_{12-18}$ primer (500 ng/μL).
3. Incubate for 10 min at 72°C.
4. Incubate at room temperature for 30 min.
5. On ice, add the following reagents (in the order indicated), and mix by gentle pipetting:
   8 μL 5X 1st Buffer
   4 μL 0.1 *M* DTT
   4 μL dNTPs (except dCTP; 0.5 m*M* each)
   1 μL RNasin (40 U)
   2 μL SuperScript II reverse transcriptase (400 U)
   5 μL [α^{33P}] dCTP dilution (50 μCi total).
6. Incubate for 1 h at 42°C.
7. Add 2 μL of SuperScript II reverse transcriptase (400 U).
8. Incubate for 1 h at 42°C.
9. Stop the reaction by adding 5 μL of 0.5 *M* EDTA and 10 μL of 0.1 *M* NaOH, and mix thoroughly by pipetting.
10. Incubate for 30 min at 65°C.
11. Neutralize the reaction by adding 25 μL of 1 *M* Tris-HCl (pH 7.5).
12. Prepare MicroSpin G-50 columns (one for each probe) or G-25 columns (two for each probe) by vortexing them briefly, opening lids, and breaking the bottoms of the columns. Follow this with a 1-min centrifugation at 3000 rpm.
13. Remove unincorporated nucleotides from the labeled cDNA probes by filtering these through the MicroSpin columns by a 2-min centrifugation at 3000 rpm.

14. Add 1 μL of the filtered probe to the scintillation cuvet with 10 mL of scintillation liquid. Measure cpm values using a scintillation counter.
15. Use total cpm values to evaluate the efficiency of labeling. Additionally, cpm values can be used to normalize probe volumes so that the number of total counts will be equal in parallel experiments. The volume of hybridization mix could also be adjusted to compensate for low cpm values of the probes.

Hybridization

1. Denature Cot1 DNA (1 mg/mL), PolyA RNA (50 μg/μL), and labeled probes by boiling these for 2 min. Chill on ice for 2–3 min.
2. Mix the required volume (10 mL or less) of fresh MycroHyb solution pre-warmed to 42°C with 250 μL of denatured Cot1 DNA (1 mg/mL), 5 μL of denatured PolyA RNA (50 μg/μL), and the required volume of the probe.
3. Discard pre-hybridization mix from the hybridization chamber; replace it with fresh MicroHyb solution mixed with denatured probe/Cot1/PolyA.
4. Hybridize the probe with microarray membranes in a rolling hybridization chamber at 42°C for 18 h.

Post-hybridization washes

1. Pre-warm washing solutions (2X SSC + 0.1% SDS, 1X SSC + 0.1% SDS, and 0.8X SSC + 0.1% SDS) to 50°C for at least 30 min.
2. Discard hybridization mix.
3. Working quickly, wash the chamber four times with small volumes (5 mL) of 2X SSC + 0.1% SDS solution pre-warmed to 50°C.
4. Change hybridization hood settings to 50°C.
5. Wash the hybridized membranes for 15 min with 25 mL volume of 2X SSC + 0.1% SDS solution pre-warmed to 50°C.
6. Again, wash hybridized membranes for 15 min with 25 mL volume of 2X SSC + 0.1% SDS solution pre-warmed to 50°C.
7. Using forceps, take the membrane from the hybridization chamber, place it over the piece of plastic wrap over the flat surface. Check membrane radioactivity by Geiger counter.
8. Wash the hybridized membranes for 15 min with 25 mL volume of 1X SSC + 0.1% SDS solution pre-warmed to 50°C.

9. Check membrane radioactivity again.
10. Again, wash the hybridized membranes for 15 min with 25 mL volume of 1X SSC + 0.1% SDS solution pre-warmed to 50°C.
11. Check the membrane radioactivity again. If necessary, wash the hybridized membranes for 15 min with 25 mL volume of 0.8X SSC + 0.1% SDS solution pre-warmed to 50°C, up to two times.
12. Immediately place the hybridized membrane (with the printed side facing up) over the plastic wrap-covered fragment of a hard plastic or phosphor screen mirror. Blot small fragments of clean Kimwipe tissue with a small volume of 2X SSC + 0.1% SDS solution, and place these next to the membrane to prevent parching. Collect the excess of washing solution with dry filter paper.
13. Cover the membrane and wet paper with a sheet of plastic wrap and brush it gently with exaggerated movements to eliminate all bubbles. Carefully attach plastic wrap to the "background" piece of plastic using a paper tape. Quality assembling may be assured by creating some tension in the surface layer of the plastic wrap cover.
14. Place an assembled "*sandwich*" containing the hybridized membrane in the exposure chamber, face up; cover it with the storage phosphor screen facing down. Place a few paper towels atop the screen to ensure good contact with the membrane.
15. Lock the exposure chamber and label it with the date of the hybridization.
16. Expose the hybridized membrane to the storage phosphor screen for 5 d at room temperature.

Data acquisition and analysis

1. Scan the exposed storage phosphor screen using phosphoimaging equipment (e.g., Molecular Dynamics STORM PhosphorImager) at high-resolution settings (50 μm/pixel).
2. Format and analyze the microarray image using available software packages for formatting, and ImageQuant (Amersham) or ArrayPro for data acquisition and background subtraction.
3. If necessary, hybridized membrane (stored wet in plastic wrap) can be re-exposed to erased storage phosphor screen, and exposure time should be adjusted according to the half-life (25.4 d for ^{33}P) and a reference (dispatch) date of radioactive isotope used for probe labeling.

Oligonucleotide Microarray Hybridization Protocol, Fluorescent Labeling

Synthesis of Fluorescent Probe

1. Pre-warm 5X 1st Strand Buffer for 3–4 min at 65°C; vortex briefly, spin briefly; keep at room temperature until used.
2. Dilute or concentrate total RNA sample (50–500 ng) to 10.3 μL volume in a 0.2- mL tube.
3. Add 1.2 μL of T7 promoter primer.
4. Incubate for 10 min at 65°C.
5. Chill on ice for 5 min.
6. Immediately prior to use, mix the following reagents (in the order indicated) in a clean tube by gentle pipetting (volumes per sample):

   4 μL 5X 1st Buffer

   2 μL 0.1 *M* DTT

   1 μL 10 m*M* dNTP mix

   1 μL MMLV RT

   0.5 μL RNaseOUT
7. To each sample tube, add 8.5 μL of the mix.
8. Incubate for 2 h at 40°C.
9. Inactivate MMLV RT by incubating for 15 min at 65°C.
10. Chill on ice for 5 min.
11. Spin samples briefly in a microcentrifuge to drive condensation off of the tube walls.
12. To each sample tube, add either 2.4 μL 10 m*M* cyanine 3-CTP (Cy3) or 2.4 μL 10 m*M* cyanine 5-CTP (Cy5).
13. Pre-warm 50% PEG solution for 1 min at 40°C, vortex briefly, and spin briefly. Keep at room temperature until used.
14. Immediately prior to use, mix the following reagents (at the order indicated) in a clean tube by gentle pipetting (volumes per sample):

    15.3 μL Nuclease-free water

    20 μL 4X Transcription buffer

    6 μL 0.1 *M* DTT

    8 μL NTP mix

    6.4 μL 50% PEG

    0.5 μL RNaseOUT

    0.6 μL Inorganic Pyrophosphatase

    0.8 μL T7 RNA Polymerase

15. To each sample tube, add 57.6 !IL of the mix.
16. Incubate for 2 h at 40°C.

Purification of fluorescent probe

1. Transfer reaction products to a 1.5-mL tube and add 20 μL of nuclease-free water.
2. Add 350 μL of Buffer RLT and mix thoroughly by pipetting.
3. Add 250 μL of 96–100% ethanol (room temperature), mix thoroughly by pipetting.
4. Transfer 700 μL of cRNA sample to a labeled RNeasy mini column placed in a 2-mL collection tube.
5. Centrifuge for 30 s at 13,000 rpm at 4°C (use a refrigerated centrifuge: Eppendorf 5417R, or similar) and discard the flow-through and collection tube.
6. Transfer the RNeasy mini column to a new 2-mL collection tube. Add 500 μL of Buffer RPE to the column.
7. Centrifuge for 30 s at 13,000 rpm at 4°C and discard the flow-through and reuse the collection tube.
8. Again, add 500 μL of Buffer RPE to the column.
9. Centrifuge for 1 min at 13,000 rpm at 4°C. Discard the flow-through and collection tube.
10. Transfer the RNeasy mini column to new 1.5-mL collection tube. Add 30 μL of nuclease-free water directly onto the RNeasy mini column filter membrane.
11. Incubate for 1 min at room temperature.
12. Centrifuge for 30 s at 13,000 rpm at 4°C.
13. Using the same 1.5-mL collection tube, add 30 μL of nuclease-free water directly onto the RNeasy mini column filter membrane.
14. Incubate for 1 min at room temperature.
15. Centrifuge for 30 s at 13,000 rpm at 4°C.
16. Mix the samples carefully (total volume should be approx 60 μL). Keeping the samples on ice, measure fluorescent dye incorporation using nuclease-free water as blank. Store at -80°C in a lightproof container.

Pre-hybridization

1. Pre-warm the Pre-Soak Solution, Pre-Hybridization Solution (Corning), and wash containers to 42°C for at least 30 min.
2. Boil dH_2O in a wide beaker and set heating plate to 100°C.

3. Holding the slide with fingers carefully and keeping the straw at approx 10 cm distance, and blow on each microarray glass slide with a compressed air.
4. Holding the slide with fingers carefully (with the printed side face down), and keep it over the boiling water until condensation covers the slide.
5. Quickly put the slide over the heating plate, pre-heated to 100°C, with the printed side face up. Immediately after the condensation disappears, remove the slide from the heating plate.
6. For crosslinking, place the slides to the turntable of the ultraviolet oven with the printed side facing up. Set energy to 800 mJ/cm^2.
7. Using a measuring cylinder, fill the wash container with 100 mL of Pre-Soak Solution. Add sodium borohydride Pre-Soak Tablet to the container and place it to a water bath pre-warmed to 42°C.
8. Allow the tablet to dissolve completely for 3–4 min, immerse microarray slides, and incubate wash container for 20 min at 42°C.
9. Transfer slides to Wash Solution 2 and incubate for 30 s at room temperature.
10. Transfer slides to another wash container filled with Wash Solution 2 and incubate for 30 s at room temperature.
11. Transfer slides to Pre-Hybridization Solution pre-warmed to 42°C and incubate for 15 min at 42°C.
12. Transfer slides to Wash Solution 2 and incubate for 1 min at room temperature.
13. Transfer slides to Wash Solution 3 and incubate for 30 s at room temperature.
14. Transfer slides to another wash container filled with Wash Solution 3 and incubate for 30 s at room temperature.
15. Transfer slides to nuclease-free water.
16. Using adapters (CombiSlide glass slide/microarray adapter, or similar), immediately dry slides by centrifuging them at 1200*g* for 3 min in a general purpose centrifuge. In this step, slides could be carefully handled by the edges.
17. Store slides in a closed container until ready to use.

Hybridization

1. Prepare glass coverslips by immersing them completely in nuclease-free water, then in isopropanol, followed by a 2-min centrifugation at 500*g* in a general purpose centrifuge.

2. Pool the appropriate volumes of Cy3 and Cy5-labeled samples (e.g., experimental and reference; 20:15 pmol or 20:10 pmol, respectively) in a 0.2-mL tube.
3. Using a concentrator, vacuum-dry pooled sample (may take 20–30 min, depending on the initial volume of samples). Samples can be covered with foil and stored at –20°C until ready to use.
4. Dissolve dried Cy3/Cy5 sample in 50 μL of Hybridization Solution (Corning).
5. Incubate for 5 min at 65°C.
6. Centrifuge for 5 min at 10,000*g*.
7. Clean hybridization chambers (Corning) by washing them with Kimwipe tissue sprayed first with RNase AWAY, then with 70% ethanol. Dry hybridization chambers with clean Kimwipe tissue.
8. To prevent dehydration, fill both wells in the base of hybridization chamber with 12 μL of nuclease-free water.
9. Blow on coverslip with compressed air; place it over the flat surface.
10. Carefully distribute 45 μL of centrifuged samples in droplets over the surface of the coverslip. Minor bubbles in the droplets could be extracted with a dry pipet tip.
11. Blow on microarray glass slide with compressed air and carefully place it over the coverslip with the printed side facing down. Invert the array and adjust a position of the coverslip, if necessary. The sample should fill the space between the coverslip and the array slide completely.
12. Without inverting, carefully place the microarray slide into the hybridization chamber, and assemble the chamber.
13. Hybridize the probe with the microarray slides in the hybridization chamber in a 42°C water bath for 14–20 h; protect samples from light during hybridization.

Post-hybridization washes

1. Pre-warm Wash Solution 1 and wash containers at 42°C for at least 30 min; set heating plate to 50°C.
2. Place a Petri dish over the heating plate pre-warmed to 50°C and fill it with 25 mL of Wash Solution 1 pre-warmed to 42°C.
3. Take the hybridization chamber out of the water bath and dry it with paper tissue. Without inverting, disassemble the hybridization chamber and place the array slide into the Petri dish with the coverslip facing up, submerging the slide completely.

4. Incubate over the heating plate (pre-warmed to 50°C) until the coverslip freely moves away from the slide (~5 min).
5. Carefully remove coverslip from the array, transfer slides to wash container filled with Wash Solution 1, and incubate for 5 min at 42°C.
6. Transfer slides to Wash Solution 2 and incubate for 10 min at room temperature.
7. Transfer slides to Wash Solution 3 and incubate for 2 min at room temperature.
8. Transfer slides to another container filled with Wash Solution 3 and incubate for 2 min at room temperature.
9. Transfer slides to another container filled with Wash Solution 3 and incubate for 2 min at room temperature.
10. Immediately transfer slides to the centrifuge. Using adapters, dry slides by centrifuging them for 1200*g* for 2 min.
11. Store slides in a closed lightproof container until ready to scan.

Data acquisition and analysis

1. Scan array slides after blowing them with compressed air.
2. Format image using available software packages; GenePix Pro software is suggested for data acquisition and background subtraction.
3. If necessary, array slides can be re-scanned within 1–2 d, with a general decrease in the recorded fluorescent signal intensity. Store in a lightproof container.

Notes

1. The use of radioactive labeling is subject to strict regulations in the context of safety concerns. All aspects of radioactive isotope application (storage, handling, liquid and solid radioactive waste disposal, etc.) should be addressed using local rules and regulations. Always be precise and careful when working with radioactive isotopes and radioactive labels. Use only dedicated sets of pipet, especially marked centrifuge, water bath, pipet tips, etc. Always work in a designated space ("*hot zone*") using personal protective equipment (shield, gloves, etc.). Check the radioactivity before, during, and after work.
2. Cyanine 3-CTP (Cy3) and Cyanine 5-CTP (Cy5) fluorescent dyes are highly light- sensitive. Minimize light exposure at every stage of the protocol, i.e., do not use direct light when handling samples, close water baths with lids, cover samples with foil when stored and handled, etc. Additionally, Cy3 and, especially, Cy5 are

sensitive to oxidation (caused by the environmental ozone) and humidity. Using the means to prevent air-drying of the slides (automated hybridization stations, where the slide is dried with pure nitrogen; Tecan) would significantly reduce the exposure to ozone. Otherwise, dye effects should be controlled for as part of the experimental design.

3. Be careful when handling microarray membranes: use long forceps with flat branches or with branches protected with fragments of plastic tubing. Glass slides could be handled using short forceps. Use only marginal areas of membranes/ slides for handling, rather than internal areas with spotted cDNA.
4. Starting from the rehydration (pre-soak) step, make sure microarray membranes/ slides are always wet. Failure to prevent drying of membrane/slide surface at any step will result in a drastic increase of background. Pay particular attention to the condition of membranes on the stages where hybridization chambers are heated.
5. Avoid multiple freezing/thawing cycles when working with DTT. Generally, any DTT aliquot should not be frozen and thawed more than five times.
6. Resulting cpm values could be used for general normalization of the required probe volumes. For example, if cpm value is 500,000 for Probe 1 and 350,000 for Probe 2, 56- and 80-μL volumes of Probes 1 and 2, respectively, should be used for hybridization. Volume of hybridization mix could be adjusted (e.g., decreased) to obtain equal final probe concentration (cpm/mL) for independent hybridizations.
7. Be sure the Geiger counter is charged and calibrated, protect the detector (probe) head with a piece of plastic wrap. Holding the probe horizontally over the hybridized membrane at approx 1 cm distance, check the approximate radioactivity of membranes after each wash. Hybridyzed membranes showing a Geiger counter reading of approx 10–15 cps do not require further washes and could be exposed to storage phosphor screen.
8. The storage phosphor screen should be checked for residual radioactivity, washed with wet Kimwipe tissue, air-dried, and subjected to erasing both immediately prior to exposure to hybridized membrane and after the scanning.
9. Modern scanners might have spatial resolution range of 5–50 to 50–250 μm/pixel. Resolution of 25–50 μm/pixel is appropriate for most readioactively labeled cDNA microarrays.

10. Low RNA Input Fluorescent Linear Amplification Kit (Agilent Technologies) contains two vials of dNTPs, labeled as "10 m*M* dNTP mix" and "NTP Mix."
11. It is suggested that both stages of Buffer RPE treatment be fulfilled, followed by the drying of RNeasy column silica-gel membrane by the extended centrifugation. Optionally, discard the flow-through and collection tube after step 7, transfer RNeasy mini column to new 2-mL collection tube and centrifuge it at 13,000 rpm for 1 min without closing the tube lid.
12. The amount of fluorescent dye incorporated could be calculated using the following formula: pmol Cy3 = $OD_{550} \times 100/0.15$; pmol Cy5 = $OD_{650} \times 100/0.25$. Using Low RNA Input Fluorescent Linear Amplification Kit, fluorescent dye incorporation in the range of 3.5–4.5 pmol per reaction can normally be obtained, depending on the amount of staring material.
13. Not more than six microarray slides should be processed in 100-mL wash container on all pre-soak/pre-hybridization steps. Use fresh solutions and clean wash containers on every step if more than six slides are processed.
14. Sodium borohydride ($NaBH_4$, Borol) is highly flammable and highly toxic when wet. Work with extreme caution in a chemical hood only; use protective equipment. Dispose sodium borohydride waste according to local regulations. Do not add sodium borohydride Pre-Soak Tablet to Pre-Soak Solution more than 15 min before use. If liquid sodium borohydride is used, mix 99 mL of Pre-Soak Solution (pre-warmed to 42°C) and 1 mL of sodium borohydride solution, immerse microarray slides immediately.
15. Some coverslips may easily be broken during centrifugation and handling. Always make some excessive coverslips in advance.
16. Not more than three microarray slides should be processed in 100 ml wash container on all post-hybridization steps. Use fresh solutions and clean wash containers on every step if more than three slides are processed.

8

CELLULAR AGING

Hammerhead ribozyme, found in economically devastating plant viroids, is one of the smallest known "*RNA scissors.*" It consists of two domains: the substrate binding region, and the catalytic core. As a result of its intrinsic catalytic property, scission of the gene target at *NUX* triplex (*N* is any base; *X* = A, C, or U) can readily be achieved in vitro. However, in biological systems, many researchers experience much lower success rates for in vivo gene silencing. This problem has contributed to the decline in the popularity of first-generation ribozymes as a major postgenomic molecular tool. However, there are no doubts that ribozyme-mediated gene silencing will stay as a very useful molecular tool for the following reasons:

1. It does not elicit interferon response in mammalian cells.
2. It possesses high specificity, avoiding "*off-targets.*"
3. It is a flexible system in terms of structural design and, thus, capable of possessing specialized features, e.g., an allosterically controllable "*intelligent*" sensor arm incorporated into the maxizymes.
4. It can cleave sequences without the need for enzymes.
5. It can function in biological, nonideal, and even industrial conditions, such as the flexizyme.
6. It can silence genes without a prior knowledge of the target sequence and is thus suited for gene discovery as with the randomized ribozymes.
7. It has potential for clinical use, e.g., an anti-HIV-1 ribozyme and anti-VEGF ribozyme (Angiozyme) are completing phase I and II human trials, respectively.

What, then, are the "ideals" that we look for in a ribozyme that would best serve in vivo gene knock-down endeavors and would strengthen its merit as an RNA therapeutic agent? Although ribozyme technology is plagued with technical pitfalls, some of these salient features and ramifications are presented. These innovations have been used in several gene knock-down experiments relevant for biogerontologists. Another utility from this next-generation ribozymes is in discovering genes involved in a particular phenotype using randomized ribozyme libraries. By inhibiting the expression of a particular gene using a large diversity of ribozymes, one can identify functional genes involved in a given pathway. Indeed, this strategy has been used by our group to identify novel gene functions for various cellular phenotypes, i.e., metastasis, apoptosis, and differentiation.

MATERIALS

Construction of Hammerhead-Ribozyme Expression Plasmids

1. tRNAVal-expression vectors:
   (a) pKE-PUR (For conventional ribozyme): as described by Koseki and co-workers, this pUC19-based plasmid contains the promoter for the human tRNA^{Val}, inserted between the *Eco*RI and *Sal*I restriction sites. The ribozyme is inserted between the *Kpn*I and *Csp45*I of the linker region.
   (b) pKE-PUR(A) (for hybrid poly(A)-ribozyme): ss described by Kato et al. by inserting 60 nt-poly(A) track downstream the *Kpn*I restriction site of the pKE-PUR vector.
2. Template DNAs:
   (a) Gene-specific ribozymes: 5^2 TCC CCG G (ttc gaa)$_{Csp45I}$↓ ACC GGG CAC TAC AAA AAC CAA CTT T(X_9) CTG ATG AGG CCG AAA GGC CGA A (X_9) (ggt acc)$_{KpnI}$ CCG↓GAT ATC TTT TTT T 3' ; where X_9 is the 9-bp gene-targeting sequence, bold letters correspond to the sequence of the hammerhead ribozyme, letters within the parentheses are the two restriction sites (indicated by the arrows), and the underlined sequence is the tRNA^{Val}.
   (b) Randomized hybrid ribozyme: 5' CCG G (ttc gaa)$_{Csp45I}$↓ACC GGG CAC TAC AAA AAC CAA CAA *NNNNNNNNN* CTG ATG AGG CCG AAA GGC CGA AA *NNNNNNNN* A (ggt acc)$_{KpnI}$CCG↓GAT ATC TTT TTT T 3' ; the randomized-binding arms are composed of the nucleotides designated by *N*.

3. Primers (10 μM):
   (a) Specific gene-targeting ribozymes: Primer A: 5' TCC CCG GTT CGA AAC 3' Primer B: 5' AGA AAA AAA GAT ATC CGG GGT ACC 3'.
   (b) Randomized hybrid ribozyme library: Primer X: 5' CCG GTT CGA AAC CGG GCA C 3' Primer Y: 5' AAA AAA AGA TAT CCG GGG TAC CT 3'.
4. *Taq* DNA polymerase (5 U/μL) (TAKARA).
5. Restriction enzymes: *Csp*45I (8 U/μL) and *Kpn*I (10 U/μL).
6. 10x *Taq* polymerase buffer: 100 m*M* Tris-HCl, pH 8.3, 15 m*M* $MgCl_2$, 500 m*M*KCl.
7. TE buffer: 10 m*M* Tris-HCl, pH 8.0; 1 m*M* ethylenediaminetetra-acetic acid (EDTA).
8. 10X L (low-salt) buffer: 100 m*M* Tris-HCl, pH 7.5, 100 m*M* $MgCl_2$, 10 m*M* dithiothreitol (DTT).
9. 50X TAE buffer: 2 *M* Tris-acetate, 0.05 *M*EDTA, pH 8.0.
10. dNTP mix (2.5 μM each of dATP, dCTP, dTTP, and dGTP).
11. Puromycin (Sigma).
12. Ethanol: 99.9% and 70% (v/v) in distilled water.
13. Phenol and chloroform.
14. Agarose.
15. DNA purification kit from agarose gels.
16. DNA ligation kit.
17. Competent *Escherichia coli* cells (JM109, DH-5α).
18. Luria-Bertani (LB) medium: 10 g tryptone, 5 g yeast extract, 10 g NaCl in 1 L distilled water.
19. LB amp plates (keep at 4°C).
20. LB broth with amp (keep at 4°C).
21. Ampicillin (10 mg/mL, filter-sterilized and kept in freezer).

Analysis of Activities of Ribozymes In Vivo

Mammalian cell culture and transfections

1. Cell lines (e.g., HT1080, MCF7, COS7)
2. Dulbecco's modified Eagle's medium (DMEM) supplemented with 10% heat-inactivated fetal bovine serum HyClone.
3. Solution of trypsin (0.25%) and ethylenediamine tetraacetic acid (EDTA) (1 m*M*) from Life Technologies.
4. Antibiotic mixture (e.g., Antibiotic-antimyoctic, Gibco/BRL).

5. Transfection reagent (FuGENE 6, Roche Applied Science).
6. Serum-free medium (e.g., OptiMEM, Gibco/BRL).
7. PBS buffer: 8 g NaCl, 0.2 g KCl, 1.15 g Na_2HPO_4, 0.2 g Na_2HPO_4 in 1 L distilled water; alternatively, we purchase 10X PBS (calcium and magnesium-free from Gibco/BRL).

Luciferase reporter assay

1. Suitable exogenous reporter plasmids for assay.
2. Cell lysis buffer (e.g., Passive lysis buffer, Promega, Madison, WI).
3. Luciferase assay kit (e.g., PicaGene Kit, Tokyo, Japan).
4. Opaque-bottom 96-well plates.

RNA isolation, northern blotting and reverse-transcription PCR

1. RNA extraction reagent (e.g., ISOGEN reagent).
2. Diethyl pyrocarbonate (DEPC)-treated water: add DEPC to deionized water at 1:1000 dilutions. Let it sit overnight at room temp and autoclave the next day.
3. 10X RNA loading buffer: 50% glycerol, 10 m*M* EDTA, 0.25% (w/v) bromophenol blue, 0.25% (w/v) xylene cyanol. Use DEPC water and keep at 4°C.
4. Hybridization buffer (e.g., ultrasensitive hybridization buffer, Ambion, Austin, TX).
5. Digitonin lysis buffer: 50 m*M* HEPES/KOPH, pH 7.5, 10 m*M* potassium acetate, 8 m*M* $MgCl_2$, 2 m*M* EGTA, and 50 μg/mL digitonin.
6. 10X C T4 polynucleotide kinase (PNK) buffer: 500 m*M*Tris-HCl, pH 8.0, 100 m*M* $MgCl_2$, 50 m*M* DTT.
7. 20X MOPS buffer: 0.4 *M* MOPS, pH 7.0, 100 m*M* sodium acetate, 20 m*M* EDTA.
8. 2X SSPE buffer with 0.1% sodium dodecyl sulfate (SDS): 1.5 *M* NaCl, 17.3 m*M* NaH_2PO_4, 2.5 m*M* EDTA, 0.1% SDS.
9. Nylon membrane (e.g., Hybond-N nylon membrane).
10. γP^{32}-ATP (10 mCi/mL).
11. Formaldehyde: 12.3 *M* in deionized water.
12. Ethidium bromide, 200 μg/mL (carcinogenic, handle with care).
13. T4 PNK (10 U/μL).
14. Reverse transcriptase (e.g., Moloney murine leukemia virus reverse transcriptase from Stratagene, Austin, TX).

Selection of Active Ribozymes Involved in a Particular Phenotype

Retroviral gene-delivery and phenotype selection

1. PLAT-E packaging cells (for mouse cells) and PT-67 (for human cells).
2. C2C12 myocytes for gene discovery for muscle differentiation.
3. pMX-puro/Rz library.
4. Polybrene (hexamidimethrine bromide).
5. Sterile ring for colony isolation.

TA Cloning and sequencing

1. pGEM-T kit.
2. IPTG.
3. T7 and M13R primers.
4. BigDye Terminator v1.1 Cycle Sequencing Kit.

METHODS

Predicting the Secondary Structure of RNA and Selecting the Ribozyme Target Sites

Given the rate-limiting step of a ribozyme cleavage reaction is its association or annealing to the target RNA, predicting the effective oligonucleotide binding sites is a major concern. Hybrid ribozymes are proved to be effective as they may overcome the limits posed by substrate accessibility.

The use of combinatorial oligonucleotide arrays, RNAseH cleavage assay, and ribozyme libraries are among the sophisticated experimental approaches that aid in determining the optimal target sites by (conventional) ribozymes. We recommend simply inspecting RNA accessibility with RNA-folding free wares, such as Mfold. However, one should keep in mind that the computer-aided approach would not account for the higher-order structures of RNAs.

1. Predict the secondary structures of the RNA target using the using the Mfold program.
2. Search for the *NUX* triplex (*N* is any base; *X* = A, C, or U), if possible, within 500 bases downstream of the start codon.
3. Determine whether these *NUX* triplexes are in extended single-stranded regions (e.g., loop regions), or unstructured areas that have at least an approx 60% open structure. Avoid stable stem structures as these are likely to be inaccessible to ribozymes.
4. Select candidate target sites.

Construction of the polIII–Driven Hammerhead–Ribozyme Expression Plasmids

Designing the ribozyme

1. Identify the 9-bp gene-targeting sequence complementary to the RNA substrate that contains the chosen *NUX* triplex.
2. Construct the hybrid ribozyme by replacing the X_9 portion with the 9-bp sequence of interest from the sequence.
3. Predict the resultant structure.

Preparation of DNA insert

1. Combine the following in a PCR reaction mix: 1 μL (0.05 μmol template DNA); 1 μL each of 100 μ*M* primer A and B; 10 μL 10x *Taq* polymerase buffer; 10 μL dNTP mix; 0.5 μL *Taq* polymerase; and 76.5 μL distilled water.
2. Execute PCR for 25 cycles of 94°C for 30 s, 58°C for 30 s, and 72°C for 45 s.
3. Check the products by running in 2.0% agarose mini-gel with 1X TAE and stained with ethidium bromide.
4. Purify the PCR product by phenol-chloroform and ethanol precipitation.
5. Perform restriction digestion of the PCR product: DNA (1 μg), 10X L buffer (10 μL), *Csp*45I (2 U), *Kpn*I (2 U), and enough distilled water to make 100 μL. Incubate the mix for 2 h at 37°C.
6. Extract digested fragments by phenol-chloroform with ethanol-precipitation. Re-suspend the digested insert in TE buffer.

Preparation of the expression vector

1. Digest 2 μg of the vector pKE-PUR with 5 U each of *Csp*45I and *Kpn*I in 1X L buffer (total volume of mixture of 100 ml) for 2 h at 37°C.
2. Check for the vector digest by running in a 1.0% agarose gel.
3. Under ultraviolet (UV) light, quickly excise the portion of the gel containing the fragment of the cut vector.
4. Purify the digested vector using a gel extraction kit (e.g. QIAGEN).
5. Mix the digested vector (0.03 pmol) and insert (0.1–0.3 pmol) to a final volume of 5 μL. Add 5 μL of solution I of the DNA Ligation kit Ver. 2 (Takara) and incubate for at least 4 h at 16°C.
6. Transform competent *E. coli* cells with the ligated mix by incubation for 30 min on ice, followed by 42°C heat shock (90 s),

and recovery (on ice, 2 min), followed by addition of SOC medium and incubation at 37°C for 40 min in a shaker.

7. Plate the cells on LB agar containing 100 μg/mL and incubate the plates for 12–16 h, or until colonies become apparent.
8. To check for the presence of plasmid, perform colony PCR. Using sterile tooth-picks or pipette tip, pick up at least five colonies to screen for the plasmid, and dip the bacteria in 6 μL distilled water (in PCR tube), and then prepare duplicate streaks on a separate LB agar plate.
9. To the PCR tube, add 1 μL 10X *Taq* polymerase buffer; 1 mL dNTP mix; 0.5 μL each of 10 μL M13 primers P7 and P8, and 1 U of *Taq* polymerase.
10. Execute PCR amplification with the following program: 25 cycles of 94°C for 30 s, 58°C for 30 s, and 72°C for 1 min. Use a negative clone (untransformed) as control.
11. Run the PCR product in 2.0% agarose mini-gel and visualize the DNA with ethidium bromide.
12. Perform mini-prep isolation of DNA from the positive clones and check the nucleotide sequence of the construct.

Analysis of Ribozyme Expression and Activity In Vivo

To analyze the effect of ribozyme on cells, either transient or stable transfection may be employed. Transient transfection is suited to quickly determine whether the ribozyme is effective in vivo using suitable assay systems, such as the luciferase reporter assay.

We can perform transient transfection if the desired phenotype is detectable within 96 h. Alternatively, stable transfection, from either bulk or clonal selection, using appropriate selection method is suited to determine the phenotypes, such as steady-state levels of the mRNA (e.g., Northern blotting) or protein (e.g., Western blotting).

Confirmation of ribozyme activity by a luciferase reporter assay

The luciferase reporter assay systems are currently one of the best nontoxic, rapid and sensitive methods for measuring the silencing effect of ribozymes on gene expression.

The basic steps for the use of the luciferase reporter gene system are as follows: (a) construction of an appropriate luciferase reporter vector, (b) transfection of the plasmid DNA into cells, (c) preparation of cell extracts, and (d) measurement of the extracts for luciferase activity.

1. Seed cells in 12-well plates and grow until 80% confluent.
2. Transfect with LipofectAMINE PLUS the following: 1-3 μg of the ribozyme-expression vector, 0.1-0.5 μg of the reporter plasmid.
3. Incubate transfected cells for 24–48 h at 37°C.
4. Add 100 μL of 1X cell lysis buffer and shake moderately for 15 min at room temperature.
5. Dispense 20 μL of the lysate into opaque-bottom 96-well plates. Be careful not to include cell clumps. Make quadruplicates for each treatment.
6. Add 100 μL of luciferin solution and immediately read light intensity in the microplate reader.

Confirmation of stable expression of ribozymes

RNA isolation

1. Aspirate medium and add 1 mL of ISOGEN reagent directly to the cells. Incubate the mixture at 4°C for 10 min.
2. Collect the cell suspension in an RNAse-free tube. Add 200 μL of chloroform to the lysate and mix vigorously for 15 s. Centrifuge the mixture at 12,000*g* for 15 min at 4°C to facilitate partition of the two solvents.
3. Transfer the (upper) aqueous phase to a new tube and add 500 μL of isopropanol. Centrifuge and remove the supernate. Add 1 μL of 70% ethanol to the pellet and centrifuge once more at 12,000*g* for 2 min at 4°C.
4. Remove the supernatant and air-dry the pellet in the hood for 5 min. Finally, add 30 mL of DEPC-treated water. Measure the concentration of RNA spectrophotometrically.

Analysis of expression level of ribozymes by northern blotting

1. Prepare the denaturing reaction mixture.
   (a) 5.5 mL of RNA (approx 2 μg) solution.
   (b) 1.0 mL of 20X MOPS electrophoresis buffer.
   (c) 3.5 ml of 12.3 M formaldehyde.
   (d) 10 ml of formamide.
   (e) 1.0 ml of ethidium bromide (200 μg/mL).
2. Incubate the RNA solution for 15 min at 65°C and then place samples on ice for 5 min.
3. Add 2 mL of 10X RNA gel-loading buffer to the sample.
4. Load samples in 2.0% agarose gel containing 2.2 *M* formaldehyde. Perform gel electrophoresis in 1X MOPS buffer.

5. Inspect quality of RNA under a UV trans-illuminator.
6. Transfer RNA to a nylon membrane (e.g., Hybond-N) for 12–16 h. Then, cross-link RNA to the membrane by UV irradiation.
7. Incubate the membrane for 1 h at 42°C in hybridization buffer.
8. During this prehybridization step, perform radio-labeling step on the probe. Mix 10X T4 PNK buffer (10 μL), ^{32}P-ATP (50 μL), and T4 PNK (10 U) in a 100-μL reaction volume a 37°C for 30 min.
9. Denature the probe by heating for 5 min at 95°C. Then, place the tube on ice.
10. Add the denatured probe directly to the prehybridization solution and incubate the setup for 12–16 h at 42°C.
11. Aspirate out the hybridization solution. Then, wash the membrane twice for 10 min in 2X SSPE with 0.1% SDS at 42°C.
12. Dry the membrane with a blotting paper and detect image under a phosphorimager.

Analysis of expression level of ribozymes by reverse-transcription PCR

1. Prior to performing the *reverse-transcription* (RT) reaction, heat 5 μg of total RNA in a 10-μL volume at 65°C for 5 to 10 min and then place on ice.
2. Set up the following components in a 1.5-mL microfuge tube: 10.0 μL heat-denatured RNA, 3.0 μL 10X PCR buffer, 2.5 μL 10 m*M* dNTPs, 6.0 μL 25 m*M* $MgCl_2$, 1.0 μL Primer D (1.8 μg/mL), 0.5 μL MMLV reverse transcriptase and 17.0 μL water.
3. Incubate the RT mix at 42°C for 1 h.
4. Denature the cDNA at 95°C and place on ice.
5. For the PCR reaction, combine the following components in a 0.5-mL PCR tube: 6.0 μL cDNA product, 1.5 μL 10X PCR buffer, 0.2 μL Taq polymerase, 0.5 μL Primer 1 (1.0 μg/μL), 0.5 μL Primer 2 (1.0 μg/μL), and 10.3 μL water.
6. Perform PCR with 20 cycles of denaturation: 30 s at 94° C; annealing: 45 s at 52°C; and extension 30 s at 72°C.
7. Run PCR products in 2–2.5% agarose and visualize with ethidium bromide.

Construction and Application of a Hammerhead Hybrid Ribozyme Library

The method for constructing a hammerhead hybrid ribozyme library is similar to that of the gene-specific hybrid ribozyme expression system described above. It differs from a specific gene-targeting ribozyme mainly in that the hybrid ribozyme catalytic core is flanked by a substrate-binding region, which is completely randomized. After restriction digestion of PCR amplified fragments, they are ligated into the plasmid, and are used for transformation of competent *E. coli* cells. However, one must intuit the maintenance of the diversity of the library, and thus some modifications in the construction of the expression plasmid are infused.

Construction of the randomized hybrid ribozyme library

1. Amplify the DNA template. Use the following PCR conditions: six to eight cycles of denaturation (95°C, 30 s), annealing (55°C, 30 s), and extension (74°C, 30 s).
2. Purify the PCR products and digest the PCR products with *Csp*45I and *Kpn*I. After cutting, again purify the DNA.
3. Ligate 550 ng of the DNA insert with 20 μg of the digested plasmid, in this case, pKE-PUR(A).
4. Use 20 μL competent *E. coli*, 1 μl of the ligation product and 180 μL of SOC for plasmid transformation. Thereafter, plate 10 (for estimating library diversity) and 100 μL (for preparing the stock) of the transformed *E. coli* into LB-amp agar. Mini-prep isolation of plasmid DNA is performed on the remaining 19 μL of cell suspension.

Transfection and selection of the desired phenotype

Following transfection of the library with commercial liposome preparations (LipofectAMINE PLUS, LipofectAMINE 2000, FuGENE 6, and so on) the cells can again be assayed as either transient or stable transfectants. The ribozymes from the selected cells are recovered and selected through, at least, two to three more cycles of both transfection and phenotype selection. If longer incubation time is required to detect a phenotypic change, it would then be necessary to create a stable line that expresses the ribozyme library. For this purpose, we recommend using retroviral or lentiviral vectors. Both possess extremely high-transfection efficiencies that are important in maintaining the ribozyme diversity.

Reports vary on strategies for selecting the cells that display salient phenotypic changes with ribozyme silencing. Immortalization-associated genes were identified after by direct picking up of foci formed after mouse fibroblasts were transduced by a hairpin ribozyme library. Similarly to the selection principle of the foci formation assay, the protocol below is given for isolating genes involved in muscle differentiation, performed previously in our laboratory, as an example of a ribozyme screening procedure for stably expressed ribozymes. A similar protocol can be used for identification of genes involved in cellular senescence by selecting population of cells that escape senescence as a consequence of ribozyme-mediated gene knock-down.

Transfection with retroviral vector and puromycin selection

1. Grow the mouse packaging cells PLAT-E to 70–90% confluency in 10-cm dish.
2. Mix 30 μL of FuGENE 6 reagent with 570 μL of OptiMEM I and incubate at room temperature for 5 min. Adequate transfection reagent should be used and according to the type of cells used.
3. Add 10 μg DNA (pMX-puro/Rz library) and leave for 15 min at room temperature.
4. Replace medium with 10 mL fresh DMEM in the PLAT-E cell plate.
5. Add the DNA–liposome complex in a dropwise manner to the cells. Swirl gently to thoroughly mix the transfection medium.
6. Incubate the cells at 37°C for 8–10 h.
7. Replace the medium with fresh DMEM and then incubate the plate at 32°C for 48 h to allow packaging of DNA into infectious viral particles.
8. Collect the culture medium containing the virus particles.
9. Filter the viral solution through a 45-μm filter and add polybrene (8 μg/mL).
10. To the 80–90% confluent rat C2C12 myocytes, replace the medium with the 10 mL viral suspension and allow for viral infection for 16 h at 37ºC.
11. Perform selection with DMEM containing 2 μg/mL puromycin for 24 h and until the cells have become 90–100% confluent.

Selection for ribozyme inhibition of muscle differentiation-phenotype

1. Incubate cells in the differentiation medium (2% horse serum in DMEM). Full in vitro muscle differentiation is achieved in parallel control cells after 8 d. Cells should be given fresh differentiation medium every 2 d.
2. Identify LPO colonies of undifferentiated cells under microscope and mark them.
3. Remove medium and wash them twice with PBS.

4. Place the "ring" over the marked colonies and add 50 μl of trypsin-EDTA.
5. After 5–7 min, remove the detach cells within the ring, transfer, and expand in a 12-well dish.

Recovery of active ribozymes

Whenever we introduce the ribozyme library to a cell population, we must consider that each transfected cell would likely to harbor more that one type of plasmid. In order to "*weed-out*" false-positive targets, we need to perform at least three to four more cycles of transfection and phenotype selection. To do this, we can combine all phenotype-altered cells and perform plasmid mini-prep to come up with a preliminary ribozyme pool. The isolated ribozymes from this first-cycle selection is then used to transform competent *E. coli* cells. A further two to three more cycles of transfection, phenotype selection, and plasmid isolation are performed with selected pool of ribozymes to reduce false-positives. Alternatively, we can follow the cloning approach. In this method, cells with the desired phenotype are individually selected and propagated. Each clone is further selected after passing through two to three cycles of phenotype selection. After streamlining our targets, perform reverse transcription PCR (RT-PCR) and sequencing, i.e., TA cloning and sequencing with T7 primer. This approach allows for greater stringency in selection and is particularly useful when only few colonies can be identified from the first screen.

RT-PCR reaction

Perform RT-PCR with primer. PCR conditions are as follows: 20 cycles of denaturation: 30 s at 94°C; annealing: 45 s at 52°C; and extension 60 s at 72°C.

TA cloning and cycle sequencing

1. Ligate the PCR product to the pGEM-T vector by setting up following the ligation reaction: pGEM-T (50 ng, 1 μL, PCR product (100 ng, 8 μL), 10 X ligation buffer (1.2 μL), T4 DNA ligase (1 μL), and distilled water (0.8 μL). Incubate the mix overnight at 4°C.
2. To prepare the IPTG plates, put LB-agar plates in an inverted position at 37°C to dry for 1–2 h to dry. Then, add 40 μL of 20 mg/mL X-gal and 8 μL of 100 mg/mL IPTG. Spread with sterile spreader and place the plates back at 37°C until use.
3. Transform chemically competent *E. coli* cells with 1 μL ligation reaction mix.

4. After overnight incubation at 37°C, put plates at 4°C for few hours until blue-colored colonies appear. Colonies containing insert will be white (or only faintly blue).
5. Check for the insert by colony PCR using T7 and M13R primers. Run PCR for 35 cycles of denaturation, 94°C for 20 s; annealing, 50°C for 45 s; and extension, 72°C for 1 min.
6. Propagate the positive clones from the LB agar streaks and perform plasmid mini-prep.
7. For cycle sequencing, mix the following: plasmid DNA (2 μL or 500 ng), 10 *μM* T7 primer (2 μL), BigDye Terminator (4 μL) and distilled water (2 μL). Run PCR for 25 cycles of denaturation, 94°C for 30 s; annealing, 50°C for 15 s; and extension, 60°C for 2 min.
8. Transfer PCR product in a 1 .5-mL tube and precipitate DNA with 80 pL of 70% ethanol.
9. Incubate on ice for 20 min.
10. Centrifuge at maximum speed for 20 min.
11. Add 100 μL 70% ethanol and centrifuge for 5 min.
12. Pipet-off excess ethanol and dry in vacuum for 10 min.
13. Add 2 μL of formamide loading buffer.
14. Store at –20°C until ready.

Identification of the candidate genes

With the information derived from sequencing the substrate-binding arms of active ribozymes in phenotype-selected cells, we can identify candidate genes either by experiment, i.e., 5' and 3' rapid amplification of DNA ends (RACE), or by simple bioinformatics. In our laboratory, we opt to screen for targets by mining the publicly accessible gene database using the Basic Local Alignment Search Tool (BLAST) program at the National Center for Biotechnology Information (NCBI) website. If overwhelmed by the number of targets, knowing many of which may be irrelevant, one may limit the search to only the ESTs (expressed sequence tags) because these are often likely targets of ribozymes.

Strategies for the Use of Ribozymes for Aging Research

The use of gene-specific and randomized ribozymes can be extended to identifying genes involved in maintenance of senescent phenotype in normal cells and detailed in the above methodology. One is likely to pick up candidate tumor suppressors or negative regulators of growth and proliferation by this protocol. Their use can be further extended to achieve immortalization of human cells by silencing such key regulators. Various assays specific to senescence cells, such as senescence-associated β-gal staining, accumulation of protein damage, and upregulation of tumor-suppressor functions, can be linked to the methodology described previously in the form of reporter assays. Furthermore, one can employ "the induction of senescence in cancer cells by various drugs" as a model system to identify novel key regulators involved in maintenance of senescence state of cells and test their role in normal aging.

Notes

1. A length of 7 bp on each side of the ribozyme may also be used: shorter arm has lower diversity, whereas a longer arm would, statistically, lead to internal secondary structures (e.g., bulges) and require more amounts of synthesis/reaction volumes to sufficiently amplify the library.
2. Other mRNA prediction free wares are also available.
3. GUC is most efficiently cleaved, followed by CUC and AUC. Other combinations may also be used.
4. Because we are assembling a new RNA, there may be drastic changes in the overall structure of the ribozyme. In the predicted secondary structure of the hybrid ribozyme, the $tRNA^{Val}$ must maintain its cloverleaf shape for nuclear export. Additionally, substrate-binding site of the ribozyme must be free and not embedded within the stem structure.
5. Adjust the settings of the UV *trans*-illuminator to the lowest intensity, if possible, to minimize damage to the DNA.
6. Avoid extending cell lysis. Dispensing of lysates and addition of luciferin should be done rapidly but carefully.
7. Work inside a clean bench to minimize RNA degradation. However, it would be ideal to have a clean bench dedicated for RNA work. Make sure that the bench is clean and the UV light is turned ON for 20–30 mins prior to RNA isolation. We directly add ISOGEN to the cells without trypsinization and washing. Cells will dislodge after adding ISOGEN and can either be collected by repeated pipetting or by scraping.
8. We recommend using an ultrasensitive hybridization buffer for overnight incubation which would have 20- to 50-fold increase in signal over traditional hybridization buffers
9. Do not use DEPC-treated water in the reaction; excess DEPCs will inhibit the RT and PCR steps.

10. The plasmids may also be electroporated because the efficiency of transformation is usually higher than that of the heat-shock method. In electroporation, however, fewer cells may be used and this would limit the diversity of the library.
11. It would be best to set the concentration of the template DNA within the range of 0.02–0.04 m*M*. Limit the PCR conditions to six to eight cycles (and not the regular 25–30). We can have sufficient amount of the insert, approx 0.2 m*M*. With more than eight PCR cycles, there is a probability that the population of ribozymes could be skewed towards the more "*amplifiable*" species.
12. A larger amount of DNA and plasmid are being used in this step to accommodate for the ribozyme diversity. Depending on the number of randomized nucleotides used, diversity is around 10^{12}, 10^9, or 10^7, when 20, 15, or 12 nucleotides are randomized, respectively.
13. Avoid the uneven replication of *E. coli* during recovery prior to selection in LB agar. Do not incubate for more than 60 min in SOC because each cell replicates at different rates. Otherwise, the population of each cell, which theoretically harbors a single type of ribozyme plasmid, would vary.
14. After 12–14 h incubation of transformed *E. coli* cells in LB-Amp plates, count the number of colonies to get an idea of the diversity of the library. If, for example, 1000 colonies grew on a plate that was initially plated with 10 mL of 200 mL cell suspension, the diversity of the library can be statistically determined: 1000 (colonies) × 200 (mL)/(10 mL) = 2 × 107. It can be noted that diversity decreased during bacterial amplification, from the theoretical 10^9 (for 15 nt-arm) down to 10^7.
15. Condition for optimization in separating the phenotype-converted cells should allow for background to be less than 1 %.
16. pGEM-T vector is compatible with JM109, DH10B, and DH5α strains.
17. We recommend plating 100 μL of cells on the LB-IPTG plates to obtain colony densities of >50 colonies per plate.
18. One μl of a two- to fourfold-diluted ligation reaction mix can be used.

Aging Cell Culture

Normal somatic cells that are serially cultured are a well studied model system that mimics the cellular and molecular changes associated with development and aging. Aging is a complex process that may consist of both environmentally stimulated and inherently programmed components. The cell culture system could be better likened to a model of cell proliferation regulatory mechanisms that shows age-associated changes. It is interesting to note that the two major age-associated diseases, cancer and atherosclerosis, demonstrate failures in cell proliferation regulation. Rubin proposes that the limit on replicative life span is a result of the inability of cells to adapt to the trauma of explantation and foreign conditions of cell culture. Rubin also addresses the doubts as to whether normal somatic cells divide only approx 50 times in an average life span in vivo. Extrapolating from data garnered from radiolabeled mouse cells and comparing it to the 70-yr life span of humans, studies have projected that human cells are able to divide between 500 and 5000 times during a lifetime (depending on cell type), and these numbers differ greatly from the 20–70 population doublings observed in vitro.

Much information has been gained in the area of therapeutic strategies, centering on the prevention and treatment of the ailments known to increase with age. Many studies have concentrated on specific age-related conditions that augment morbidity and mortality, as aging itself is more difficult to address and mimic in an in vitro culture condition. To this end, research approaches either focus on the age-associated changes with unfavorable consequences or the processes underlying many age-related dysfunctions. Questions arise as to whether such studies accurately represent what occurs inside of an organism. Cultivation requires that cells continuously undergo proliferation, which does not occur in an organism and which stresses cells unnaturally. Moreover, cells in culture and those in an intact organism differ drastically in metabolic needs, growth conditions, etc. The conventional 10-fold dilution of serum in culture media contains a lower concentration of protein than is normally found in extracellular tissue fluids. To strengthen the argument that the needs of cultured cells differ from those of cells in intact tissue, Rubin cites the study illustrating that freshly isolated cells require 13 amino acids for stable growth, whereas cells in an organism require 8–10. Many cells fail to multiply unless seeded at a relatively high population density. This fact is reiterated by the great effect on proliferative capacity of conditioned media, especially on small numbers of cells.

The limited growth potential observed in cultured somatic cells has been labeled an in vitro artifact of genetic damage caused by the synthetic environment, but a normal human cell population with infinite proliferative capacity has not been reported. Moreover, in light of the extremely low incidence of spontaneous immortalization, it is indeed probable that the replicative potential of human cells is inherently limited. Although the decrease in cell growth potential during serial cultivation has been implicated in causing aging, there is no evidence that correlates human aging with somatic cells losing their potential to divide. Furthermore, studies have failed to consistently correlate decreased telomere length, increased lysosomal enzyme function (i.e. â-galactosidase staining), or genes involved in cell cycling with senescence rather than a reversible, prolonged postmitotic state. Yeast use recombination and *Drosophila* utilize specific retrotransposons to lengthen the ends of chromosomes. The limited proliferative capacity of somatic cells appears an appropriate system by which to study human aging, as long as correct correlations are made between observed phenomena and the physiology of the entire organism. The culture system exhibits the gradual, stepwise changes that permanently alter the organism, not so much in discrete phases as in a continuum of events that mirror the time from the fusion of the gametes to the demise of the organism.

Culturing Aging Fibroblasts

Procedures for culturing aging fibroblasts

Materials

The following materials are for mammalian cells grown in a monolayer in Petri plates or flasks. Sterile conditions must be maintained at all times and all culture work should be performed under a laminar flow hood. Although it is customary to add antimicrobial agents to culture media to prevent contamination, it is not completely necessary. Indeed, prolonged usage may cause cell lines to develop antibiotic resistance or lead to cytotoxicity. Moreover, certain combinations of antimicrobial substances may be incompatible. For most aging cell types, an antibiotic/antimycotic solution of streptomycin, amphotericin B, and penicillin added to the media at a final concentration of 1% adequately hinders the growth of bacteria, fungi, and yeast. All incubations and long-term culturing must be done in a humidified 37°C, 5% CO_2 incubator. Media, trypsin/EDTA solutions, and *phosphate-buffered saline* (PBS) should be warmed to 37°C in a water bath before use. Cells are grown and subcultured in appropriate media.

1. Monolayer cultures of cells.
2. Trypsin/EDTA solution.

3. Complete medium with serum.
4. Sterile Pasteur pipets.
5. 70% (w/v) ethanol.
6. Sterile PBS without Ca^{2+} and Mg^{2+}.
7. Tissue culture plasticware (pipets, flasks, plates, cryovials, 15- and 50-mL conical tubes), all sterile.

Methods

The methods outlined as follows describe the thaw and recovery of frozen cells, the trypsinization and subcultivation of monolayer cells, and the freezing of monolayer cells.

Thaw and recovery

1. Remove cryovial from liquid nitrogen and place in a 37°C water bath. Agitate gently until thawed.
2. Wipe vial with 70% ethanol before putting under hood and opening.
3. Transfer cells to a sterile 15 mL conical tube containing medium prewarmed to 37°C. Centrifuge 5–10 min at 150–200*g* (approx 1000 rpm in a swinging bucket or 45° fixed-angle rotor). Discard supernatant containing residual dimethyl-sulfoxide (DMSO).
4. Resuspend pellet in 1 mL complete medium and transfer to culture plate/flask containing the appropriate amount of medium. Place in incubator. Check cells for adherence after 24 h.

Trypsinization and subcultivation

Cells should be subcultured upon approx 90% confluence to avoid contact inhibition or transformation.

1. Remove media from primary culture with a sterile Pasteur pipet. Wash adherent cells with a small volume of sterile PBS to remove residual *fetal bovine serum* (FBS), which may inhibit trypsin.
2. Add sufficient 37°C trypsin/EDTA solution to cover the cell monolayer. Place in incubator for 1-2 min. Check cells with an inverted microscope to make sure cells are detached.
3. Add appropriate amount (2-6 mL) of warmed culture medium and disperse cells by gently pipetting up and down.
4. Add appropriate aliquots of the cell suspension to new culture vessels.
5. Incubate cultures, checking for adherence after 24 h.
6. Renew medium every 2-4 d.

Freezing

Cells to be frozen should be in late log phase growth.

1. Dissociate monolayers with trypsin/EDTA and resuspend cells in complete medium.
2. Count resuspended cells to determine viability and number.
3. Gently pellet cells for 5 min at approx 300–350*g* (1500 rpm in a swinging bucket or 45° fixed-angle rotor) and remove and discard supernatant above the pellet.
4. Resuspend cells in freezing medium at a concentration of approx 5×10^6 cells/mL. Aliquot cell suspension into labeled cryovials and freeze immediately.
5. Cells can be frozen at –80°C for short-term storage. (Optional: place cryovials into a styrofoam container or slow freezer such as the Nalgene Cryo Freezing Container filled with isopropanol, which has a repeatable –1°C/min cooling rate to prevent the formation of damaging ice crystals.)
6. Transfer cells in cryovials to liquid nitrogen (–196°C) after 24 h for long-term storage.

Major problems and experimental considerations

Relationship of replicative senescence to in vivo senescence

One major question in the study of aging is whether the changes we observe in replicative senescence correlate with the pathways and mechanisms of cell senescence *in situ*. Support for the direct relationship between replicative senescence and cell senescence *in situ* has come from the decline in the propagative lifespan of skin fibroblasts in culture as a function of donor age. However, consistent findings confirming the inverse relationship of donor age to proliferative capacity have not been shown, which has been problematic for defining the relevance of the cell culture model to aging of the entire organism.

A major source of controversy regarding replicative senescence is the fact that many authors have modulated certain pathways and cell culture conditions leading to the senescent phenotype, showing that the senescent phenotype can be achieved despite proliferation. This evidence suggests that the senescent phenotype is a final, common pathway for actively dividing cells in which signaling and/or metabolic imbalances occur. This has led to the conclusion that cells may not be able to achieve their true differentiated fate in vivo because of signals missing from the culture medium or failure to process these signals.

However, establishing this direct relationship is not a requirement for utilizing cell culture as a tool for studying aging mechanisms. In fact, fibroblasts in culture express human genetic, metabolic, and regulatory behavior allowing the cells to undergo changes in a predictable and reproducibly constant environment.

Extra-culture conditions

Cells maintained in culture are continuously exposed to full-spectrum fluorescent light, which can be detrimental to stable cell conditions. This light can affect culture conditions as well as the cells themselves. It has been illustrated that exposure to fluorescent light can damage photoactive components of cell culture media. Also, repeated exposure to standard fluorescent light greatly enhances the frequency of single-stranded DNA breaks.

There are several other conditions that can induce a premature senescent phenotype, such as hydrogen peroxide, *tert*-butylhydroperoxide, and ultraviolet and gamma radiation. Another source of cellular stress is the trypsinization procedure utilized to remove adherent cells from the substratum. It was shown by comparing different frequencies of trypsinizations to a non- enzymatic technique that weekly trypsinizations do not affect the proliferative capacity of the mass culture.

Timeline

There are some inherent drawbacks to studying aging in vitro that should be considered when outlining a study. Aging studies are often time-consuming. For accuracy, it is better to investigate one serially cultured sample at different ages than to examine a few different cultured samples at different time periods. This often takes months of assaying young cells, growing the same culture to near-senescence, and assaying the aged cells. One solution is to start multiple cultures offset by a few days so that cells can be cultured and assayed according to their experimental design simultaneously. While serially culturing aging cells, it is important to store cells at various ages.

Freezing cells in liquid nitrogen is a good way to keep cells at a certain age. Should a problem arise with a culture of a certain age, re-freezing enables the researcher to begin the study again. This also obviates frequent re-ordering of cells at specific ages, which is contingent on their availability. The Coriell Institute provides many different cell lines used in aging research; however, not all ages are available consistently.

In Vitro Signs of Aging

Morphological signs of aging

Evidence for the correlation of aging cells in culture with cellular aging in vivo exists, especially in light of the events that occur both in vitro and in vivo. Macieira-Coelho reviews the overlapping events, which include the loss of division potential; chromatin alterations; loss of the capacity to migrate; increase in cell size, volume, and protein content; and decreased mitogenic response to growth factors.

Characteristic morphological changes accompany replicative senescence observed in culture, namely increased cell size, nuclear size, nucleolar size, number of multinucleated cells, prominent Golgi apparati, increased number of vacuoles in the endoplasmic reticulum and cytoplasm, increased numbers of cytoplasmic microfilaments, and large lysosomal bodies.

These cells also seem to exhibit an increased sensitivity to cell contact, perhaps as a result of changes in interactions with the *extra-cellular matrix* (ECM) or expression of secreted proteins, resulting in reduced harvest and saturation densities. Although the synthesis of macromolecules decreases with increasing age, the intracellular content of RNA and proteins increases.

These elevations could contribute to the increased cell and nuclear size and numbers of inclusion bodies observed in late-passage cells, and might be caused by reduced protein degradation by proteosome-mediated pathways, declined RNA turnover, and the uncoupling of cell growth from cell division (and putative block of senescent cells in late G_1).

Bio-markers of aging

Some important biological markers of aging exist that could help identify senescent cells in culture and in vivo. Some markers include IGF-1, EGF, c-fos, and senescence-associated β-galactosidase. IGF-1 is produced by many cell types and plays an important role in regulation of cell proliferation. Ferber et al. investigated the production of IGF-1 and its ligand, the IGF-1 receptor, in normal diploid fibroblasts that were both presenescent and senescent. The results illustrated that IGF-1 mRNA production was lowered to undetectable levels in senescent cells whereas IGF-1 receptor mRNA production remained at detectable levels.

Another biological marker that is differentially regulated in senescent cells is *epidermal growth factor* (EGF). EGF signaling has been postulated to be impaired in nonproliferating senescent human diploid fibroblasts downstream of receptor binding. Carlin et al. compared lysates from young and senescent WI-38 cells for proteolytic activity directed toward the EGF receptor. Their results indicate a protease that cleaves the 170 kDa EGF receptor and is produced only in senescent fibroblasts. The production of the proto-oncogene c-fos is important in growth regulation, as it is part of the AP-1 transcriptional activator. In 1990, Seshadri and Campisi demonstrated a loss of c-fos inducibility in senescent WI-38 cells, suggesting that lack of proliferation was in some part due to selective repression of c-fos. Staining for β-galactosidase is a technique to label senescent cells *in situ* and has been used in many studies to determine the amount of senescence achieved in culture. Principles and techniques for senescence-associated β-gal staining (SA-β-gal) are discussed later.

Quantification of cellular aging

The aging process in vitro is accompanied by a loss in proliferative potential, which results in a decline in cellular replication. The factors involved in this reduction are as yet unknown. During aging studies, it is necessary to detail growth and keep track of cellular age in terms of population doublings, or the number of times a cell population doubles after a predesignated period of time.

Generally, population doublings can be recorded by plating a specific number of cells and then counting those cells after a defined period of growth. If the population has doubled, for example a plating of 1.0×10^6 cells is followed by a growth phase and a count of 2.0×10^6, one population doubling is added to the current age of the cells. However, because even counts as specified above are almost never achieved, a more precise method is needed for accuracy.

Notes

1. Most cell types require a 0.25% (w/v) trypsin/0.2% (w/v) EDTA solution prepared in sterile Hank's Balanced Salt Solution (HBSS) or 0.9% (w/v) NaCl to detach cells and chelate Ca^{2+} and Mg^{2+} ions that could hinder the action of trypsin.
2. With the exception of BJ fibroblasts and other primary cell cultures, 1–2 min submerged in trypsin at room temperature or 37°C should be sufficient to disperse cells. BJs and other difficult cell types may require 5–15 min of incubation at 37°C, with occasional checks under an inverted microscope to ensure that cells are detached. To avoid clumping of cells with fibroblast-like morphology, do not agitate the cells by hitting or shaking the flask/culture vessel. Continue to place cells at 37°C until they are no longer adherent or until 15 mins have passed (extended trypsin exposure damages cells).

3. It is important to disperse clumps before counting cells with a hemacytometer. To check for viability, 0.75 mL of cells is mixed with 0.25 mL of trypan blue. Nine microliters of this suspension is loaded onto the hemacytometer with coverslip and viewed under the microscope (100× magnification). Viable cells remain unstained. After counting, calculate the number of cells per milliliter: cells/mL = average count per square of grid × dilution factor × 10^4.
4. Freezing medium is composed of 90–95% complete culture medium supplemented with 5-10% DMSO or glycerol.

9

NUCLEUS IN AGING

Meiotic division is unique in that two successive metaphases occur without an intervening interphase to produce gametes (*oocyte* or *sperm*). During the first meiotic division (MI), homologous chromosomes are segregated while two sister chromatids remain attached to each other by cohesion and move to one spindle pole, resulting in a reductional division, followed by an arrest again at the metaphase of the second meiotic division (MII) in mammalian oocytes, until fertilization. To ensure equal chromosome segregation, each homolog in MI and each sister chromatid in MII must align properly on the spindle's metaphase plate. At puberty, gonadotrophin stimulates follicle development and induces meiotic resumption and division of competent oocytes enclosed in cumulus mass, as indicated by *germinal vesicle breakdown* (GVBD) and progression to MII stage, and extrusion of a polar body with half reduction of genomic DNA. The fertilization of MII oocytes produces zygotes and initiates embryo development in mammals.

Age-related decline in fertility is a common phenomenon in women and in females from many other long-lived mammalian species. The high rate of success of egg donation, even in older women, demonstrates that poor oocyte quality is a major cause of the age-related decline in female fertility. Further, chromosome alignment and structure of the meiotic spindle in oocytes are significantly altered in older women, leading to a high prevalence of aneuploidy.

Most aneuploidies associated with maternal aging are believed to derive from non-disjunction and meiotic errors at meiosis I. Reduced chiasmata or reduced cohesion between chromatids during the prolonged prophase I arrest with increasing maternal age also may lead to premature chromatid separation, resulting in aneuploidy. Some evidence suggests that chromosomes can control meiotic spindle formation and instructions for proper spindle behavior at meiosis I reside within the chromosomes, not within the spindles. Other evidence suggests that nondisjunction and meiotic errors arise from cytoplasm. Further, maternal age affects not only meiosis, but also early embryo development.

Dysfunctional cytoplasm, particularly mitochondrial dysfunction and mutation in mitochondrial DNA (mtDNA), may increase generation of free radicals and reduce ATP levels, which, in turn, could affect chromosome segregation and/or spindle assembly. Mitochondrial dysfunction has been implicated in the pathogenesis and etiology of Down's syndrome, and animal studies support a role for mitochondria and mtDNA in the meiotic apparatus.

It is unclear how aging contributes to nondisjunction and meiotic errors, and until recently, the lack of a rodent model of oocyte aging hindered progress in this area. Senescence-accelerated mouse (SAM) resemble human females in reproductive aging, as evidenced by significantly increased frequency of metaphase chromosome misalignment and disruption of spindles. Using germinal vesicle (GV) nuclear transfer, we and others found that nuclear or nuclear-associated defects appear to be a major factor that contributes to metaphase chromosome misalignment and meiosis abnormality within reproductive aging. GV nuclear transfer itself does not influence normal meiotic progression as determined by karyotyping or development, as well as spindle structure and chromosome alignment after meiotic maturation. The carryover of cytoplasmic mitochondria also is minimal. Thus, the technique can be reasonably employed for evaluating the contributions of cytoplasmic vs nuclear factors to aberrations in meiosis with reproduction aging. The knowledge gained from these studies can be directly relevant to understanding, and potentially treating, human aging-associated oocyte infertility.

Nuclear transfer has been used as a potential tool for dissecting the contribution of nucleus vs cytoplasm to apoptosis. Indeed, apoptotic potential could be transferred cytoplasmically in mouse zygotes. Cytoplasm, probably through mitochondria, may play a central role in mediating cell cycle arrest and apoptotic cell death. Further, nuclear transfer protected genomes from telomere dysfunction and promoted cell survival by reconstitution with cytoplasm with functional mitochondria, indicating that healthy cytoplasm can rescue nuclei at the early stage of apoptosis. Study of the role of cytoplasm in the regulation of development and death of reconstituted zygotes provides a model to investigate the role of this cellular compartment, which includes mitochondria, in apoptosis during early development.

In this chapter, we first describe the techniques of nuclear transfer at the GV stage, which is utilized to understand biological mechanisms underlying aging-associated meiotic defects. Second, we describe the method for nuclear transfer in zygotes at pronuclear stage.

MATERIALS

Germinal Vesicle Nuclear Transfer

Reagents

1. Pregnant mare serum gonadotropin (PMSG).
2. Minimum essential medium (MEM) with Earle's salts and L-glutamine.
3. Cytochalasin D. Highly toxic.
4. Myo-Inositol.
5. Dimethylsulfoxide (DMSO).
6. Light mineral oil.

Medium preparation

1. In vitro maturation (IVM) medium for denuded mouse oocytes: MEM supplemented with 0.24 m*M* pyruvate and 5 % fetal bovine serum, 0.22-μm filter sterile. Pyruvate stock can be made at the concentration of 240 m*M* (1000X), aliquot, and stored at –20°C.
2. GV transfer micromanipulation medium: 25 m*M* HEPES- buffered IVM medium, supplemented with 50 μg/mL 3-isobutyl-L-methylxanthine (IBMX, I-7018, Sigma) and 2 μg/mL cytochalasin D, 0.22-μm filter sterile. IBMX stock can be made at the concentration of 50 mg/mL in DMSO (1000X), aliquot, and stored at –20°C. Cytochalasin D stock can be made at the concentration of 1 μg/μL in DMSO, aliquot, and stored at –20°C.
3. Electro-fusion medium: 0.28 *M* mannitol (M-9647, Sigma), 0.1 m*M* $CaCl_2$ (C-7902, Sigma), 0.1 m*M* $MgSO_4$ (M-2393, Sigma), and 5 m*M* histidine (or 0.025% BSA) in deionized water, 0.22-μm filter sterile.

Equipment and tools

1. Inverted microscope, equipped with Normarski interference optics, Axiovert 100TV, Zeiss.
2. Two micromanipulators, Narishige, Japan.

3. Two micro-syringes for applying suction or positive pressure, Narishige, Japan. Syringes, connecting plastic tubes, and holders are filled with Fluorinert FC-40 (Sigma), with no air bubbles in the path.
4. Microforge, MF-9, Narishige, Japan.
5. Pipet puller, Sutter P-97.
6. Micro-grinder, EG-44, Narishige, Japan.
7. Borosilicate glass capillary tubing length 10 cm. Standard wall outside diameter 1.0 mm; inside diameter 0.58 mm. Thin wall outside diameter 1.0 mm; inside diameter 0.78 mm. Warner Instrument Corp.
8. Manipulation chamber, a simple chamber can be made by gluing a glass frame (using nontoxic glass glue) onto a long thin (25X55X1) glass coverslip, which allows maximum light passing through. Droplets (10–50 μL of manipulation medium with appropriate flatness to prevent light reflection) are made on the coverslip within the chamber covered with mineral oil to prevent evaporation.
9. Dissection stereomicroscope, SMZ-1B or 2B, Nikon, Japan.
10. CO_2 incubator, Model 3130, Forma Scientific, Inc., USA.
11. BTX 2001 Electro Cell Manipulator, Genetronics, Inc., San Diego, CA.
12. Alcohol burner.
13. Two pairs of Watchman forceps.
14. One pair of surgical forceps, and one pair of small curved forceps.
15. One pair of surgical scissors, and one small curved scissors.
16. Thirty-gauge needle with 1 mL disposable syringe.
17. Thirty-five-millimeter Petri dish (351008, Falcon).
18. Sixty-millimeter Petri dish (351007, Falcon).
19. Heat-pulled transfer pipets, with opening diameter of 100–120 jim.
20. 70% ethanol.

Preparation of microtools

1. Holding pipets. Holding pipets are made from capillaries (1 mm in diameter outside) above appropriately sized burner flame by turning around and pulling using two hands, to an outside diameter of approx 50–100 μm. A shaft of 1–1.5 cm is smoothly scratched using a diamond pen on clean Kimwipe supported by a finger, then bent slightly to create a clean flat cut. The tip is polished 45° over a glass bead melt on a platinum filament using a

microforge, making inner diameter approx 15–20 μm. The pipette is reoriented and the shaft is bent about 10–15° with heated filament. The bent side is marked on the pipet for easy installation into the pipette holder.

2. Needles. Needles are produced from thick wall glass capillary tubing. The capillary tube is secured into the center of heating element in the pipet puller. The setting of the puller is selected to make the needle sharp at the tip and strong in the shaft. If this is not satisfied, the needle can be modified using the microforge.
3. Blunt nuclear transfer pipets. Transfer/enucleation pipettes are made from borosilicate glass capillary tubing (1.0 mm outer diameter [OD], 0.78 mm inner diameter [ID]). The pipet is pulled on a horizontal pipet puller by adjusting appropriate heat, puller, velocity and time, so that a thin shaft of 15–30 μm less than 0.5 mm long is created. To make the instruments, we use Narishige microforge with a ×10 objective, a platinum filament of diameter of 0.2 mm. The V- shaped filament is mounted horizontally in the microforge, and a small glass bead is fused onto the end of the filament. An even broken tip of an outside diameter of approx 15–20 μm is created by touching onto small glass bead made on a slightly heated thin platinum filament (0.1 or 0.2 mm), and instantly turning off heat using microforge. The heating on and rapid off constricts the glass and breaks the tip. The tip is slightly polished without visibly affecting opening of the tip, over a glass bead melt on a platinum filament, to smooth the tip and avoid damage to eggs during micromanipulation.
4. Beveled nuclear transfer pipets. First, similar to making blunt end pipettes, an even broken tip of outside diameter approx 15–20 μm is made using a microforge after pipette pulling. A thin-walled capillary is drawn on the pipet puller at a setting that will produce a shaft of 10–15 mm from shoulder to tip. The pipet at an outside diameter of approx 20 μm is slightly fused to the glass bead near the tip of the filament, and the heat is switched off rapidly, such that contraction of the filament and glass break the tip off squarely and cleanly on the microforge, without causing narrowing of the tip. It is important not to bend the pipet when breaking on microforge. The pipet is then attached to a micro-grinder with high speed, lowered about 45° down onto grinding wheel, while sterile distilled water is being injected onto the wheel to clean the glass dust. The tip is beveled and clean.

The pipet tip must be sharpened for penetration of the zona pellucida using the microforge. The pipet is put back to the microforge, and the ground edge is then smoothed by bringing the bevel close to the glass bead, being heated a little. The tip of the pipet is touched to a small glass bead of fairly low heated filaments, such that the tip is a little fused with the bead, then the pipette is rapidly drawn away from the bead to form a short sharp spike. Care should be taken to avoid narrowing the opening. A long, sharp spike can cause damage to membrane and lysis of oocytes or embryos. So, it is also important not to stretch the tip when pulling the spike. Finally, the shaft of the pipet is bent about 10–15° on the side of the beveled tip with heated filament. This will ensure that the beveled tip can be seen under the microscope after mounting into the pipet holder, for convenience of micromanipulation. To avoid the occurrence of sticky cytoplasm, the pipet tip can be siliconized with Sigmacot (Sigma), washed with sterile distilled water, dried, and stored in a sealed container.

Pronuclear Transfer

Reagents and solution

1. Chorionic Gonadotropin (hCG) (CG-10, Sigma).
2. PMSG (367222, Calbiochem).
3. Hyaluronidase (H-6254, Sigma).
4. Preparation of potassium simplex optimized medium (KSOM) and HEPES (14 m*M*)-buffered KSOM (HKSOM) for embryo culture and in vitro manipulation, respectively.
5. Micromanipulation medium: HEPES (14 m*M*)-buffered KSOM medium (HKSOM), supplemented with 2 μg/mL cytochalasin D and 2 μg/mL nocodazole (Sigma), 0.22-μm filter sterile. Nocodazole stock (highly toxic!) can be made at the concentration of 1 μg/μL in DMSO, aliquot, and stored at –20°C.
6. Fusion medium: 0.28 *M* mannitol, supplemented with 0.1 m*M* $CaCl_2$ and $MgSO_4$, and 5 m*M* histidine (or 0.025% BSA). 0.22-μm filter sterile.

Equipment and microtools

Essentially the same as described above.

Methods

Germinal Vesicle Nuclear Transfer

1. Collection of GV oocytes and IVM of denuded oocytes. The isolation and culture of immature oocytes are performed essentially following

procedures described. Female mice are primed by a single i.p. injection of 5 IU of PMSG 44–46 h prior to collection of GV oocytes, and euthanized humanely by cervical dislocation. Abdominal areas are sterilized with 70% ethanol. Skin and peritoneum are cut to open and expose the abdominal cavity. Ovaries are removed and ovaries with antral follicles punctured peripherally with a 30-gauge needle in 2–3 mL pre- equilibrated IVM medium in a 35-mm Petri dish, and cumulus GV oocyte-complexes are released, collected, and washed in IVM medium droplets three to four times. Cumulus cells are removed mechanically from oocytes by gentle repeated pipetting, to facilitate micromanipulation. The inner diameter of pipet used for stripping cumulus cells from GV oocytes approximates 80 μm. Caution is needed to prevent possible lysis of oocytes during pipetting. Ten to 20 GV stage oocytes are washed and cultured for 15–16 h in a 100-μL droplet of IVM medium, under mineral oil at 37°C in an atmosphere of 7% CO_2 in humidified air, for in vitro maturation. Nude oocytes without cumulus cell layers released from ovarian follicles are excluded from experiments. All in vitro manipulations are performed at 36–37°C on heated stages or chambers.

2. At 20 min before micromanipulation, GV oocytes are incubated at room temperature in micromanipulation medium.
3. GV can be removed by two methods, one using a sharpened beveled micropipet, and the other involving in cutting zona pellucida (or zona drilling) with a needle, followed by aspiration of GV with a blunt nuclear transfer pipet. Without using a needle, Piezo pipet driver and blunt pipette may also be used for GV transfer.
4. The oocyte is held and rotated such that some perivitelline space under zona pellucida can be easily seen. The microneedle is then used to penetrate the zona tangentially into the perivitelline space against the holding pipet inserted through perivitelline space. The needle is pushed forward, makes another hole in the zona, and penetrates out of the perivitelline space, with the strong shaft region of the microneedle inside the isolated zona pellucida. The distance between the two holes is about 20–30 μm in diameter.
5. Release the oocyte from the holding pipet by applying mild pressure; the oocyte remains attached to the microneedle. Using the joystick controller, move the microneedle with the isolated region of zona pellucida and rub against the outside wall of the holding pipet by moving up and down to achieve greater contact between the holding

pipet and the microneedle. Do this until a hole is created in the zona pellucida and the oocyte detaches from the microneedle. A batch of 30 or so oocytes can be done with zona drilling within 15 min. Donors and recipients can be manipulated on the same slide using microdrops to separate them.

6. Put a batch of oocytes that have zona drilling into the chamber. Use the holding pipet to rotate and suck the oocyte so that the zona opening can be easily seen at the 3 o'clock position. The blunt pipet, guided by the joystick of the micromanipulator, can easily enter the perivitelline space. The GV nucleus enclosed with oocyte plasma membrane and very little cytoplasm (karyoplast) is then drawn into the micropipet by applying gentle suction. The GV will enter the shaft of the pipett and stop suction, after which the pipet is slowly withdrawn to reseal the plasma membrane.
7. Thereafter, the isolated GV karyoplast is inserted with the same tool into the perivitelline space of another previously enucleated oocyte at the GV (GV–cytoplast). A "*grafted oocyte*" constitutes a manipulated oocyte in which an isolated karyoplast and an isolated cytoplast are still distinct entities. The resultant GV–cytoplast complexes are incubated for 15–20 min in IVM medium added with IBMX at 37°C in 7% CO_2 pending electrofusion.
8. Electro-fusion: GV–cytoplast complexes are placed in the fusion medium between platinum electrodes of a fusion chamber. They are first aligned with an AC pulse of 5 V for 4–5 s, and then fused with two DC electrical pulses of 1.8–2 kV/cm for 50 μs. The fusion procedure is carried out using a BTX 2001 Electro Cell Manipulator, and can be repeated three times, if necessary, with an interval of 30 min between pulses. The incorporation of GV nuclei into the cytoplast is monitored 30 min after each electric pulse. Fused, reconstituted oocytes are then washed and cultured in IVM medium without IBMX for in vitro maturation as described previously.

Pronuclear Transfer

Collection of zygotes

Female mice are subjected to a 14-h light/10-h dark cycle and cared for according to the procedures approved by your institution's Animal Care and Use Committee. Superovulation of females is achieved by intraperitoneal injection of 5 IU PMSG, followed 46–48 h later by injection of 5 IU human chorionic gonadotrophin (hCG). Females are then mated individually with males with proven fertility and are selected

Table 9.1. Medium for Mouse Embryo Culture (KSOM) and Manipulation (HKSOM)

Components	*KSOM stock 10X (mg/100 mL DH_2O)*	
NaCl	5550.00	
KCl	186.00	
KH_2PO_4	47.60	
$MgSO_4$	24.10	($MgSO_4 \cdot 7H_2O$) 49.30
EDTA	3.00	
D-Glucose	36.00	
Sodium pyruvate	22.00	
Sodium lactate	1.86 mL	
L-Glutamine	73.10	
Phenol Red	4.50	
Aliquot 10–11 mL tube	stored at -20°C	Not sterile
$CaCl_2 \cdot 2H_2O$ (stock 1000X)	1.255 g/5 mL DH_2O	stored 4°C
Working solution	*HKSOM (50 mL)*	*KSOM (50 mL)*
$NaHCO_3$	0.01745 g (4 m*M*)	0.105 g (25 mM)
Hepes	0.1668 g (14 m*M*)	0.033 g (2.5 mM)
Dissolved in DH_2O (Sigma)	44 mL	44.25 ml
BSA	0.050 g	0.050 g
KSOM Stock 10X	5 mL	5 mL
$CaCl_2 \cdot 2H_2O$ (stock 1000X)	50 μL	50 μL
NEAA (100X, Sigma)	0.5 mL	0.5 mL
MEM AAS (50X, Sigma)	0.5 mL	0.5 mL
Antibiotics (100X, Gibco)	0.25 mL	0.25 mL
NaOH (1 *N*)	0.24–0.25 mL (pH 7.2)	
0.4 g NaOH to 10 mL H_2O (1N)		
0.22-μm Filter sterile stored 4°C, valid for <1 mo		

for successful mating based on vaginal plugs the next morning. At 20–21 h after hCG injection, zygotes (day 0.5) enclosed in cumulus masses are released from oviduct ampullae into the modified HEPES-buffered KSOM containing 14 m*M* HEPES and 4 m*M* sodium bicarbonate (HKSOM) with 0.03% hyaluronidase, and then cumulus cells are removed by gentle pipetting. Care should be taken to minimize exposure to hyaluronidase. Cumulus-free zygotes are washed in HKSOM then in pre-equilibrated, modified KSOM, supplemented with amino acids and 2.5 m*M* HEPES. Embryos are cultured in 50-μL droplets of KSOM under mineral oil at 37°C in a humidified atmosphere of 7% CO_2 in air.

1. Before collection of zygotes, culture dishes are prepared. Eight 50-μL droplets of KSOM can be placed in a 35-mm Petri dish,

then covered with 3 mL mineral oil and incubated in the CO_2 incubator for at least 1 h or overnight.

2. Euthanize the females by cervical dislocation.
3. Replace the animal back on a clean absorbent paper, and wet the abdominal area with 70% ethanol.
4. Cut a lateral incision with the surgical scissors. Tear the skin outward toward the head and tail with two pairs of surgical forceps. Use curved scissors to cut open the peritoneum and expose the abdominal cavity. Pull intestines aside and expose uterine horns and ovaries.
5. Use forceps to grasp each uterine horn close to the oviduct and tear apart the connecting supporting tissues underneath. First, cut between the oviduct and ovary, then cut the uterine horn and place the complete oviduct in HKSOM in a 35-mm Petri dish.
6. Repeat this process for the other oviduct and for other females.
7. Make one droplet (100–200 μL) of 0.03% hyaluronidase and several droplets of HKSOM in a 60-mm petri dish. Place the isolated oviducts in the hyaluronidase droplet.
8. Under the steromicroscope, a swollen area—the ampulla—containing fertilized eggs is apparent and visible.
9. Use two pairs of watchman forceps to tear open the ampulla area of oviduct and release the fertilized eggs. The released eggs are detached from surrounding cumulus cells by the hyaluronidase.
10. Wash the zygotes by transferring them through several droplets of HKSOM with a transfer pipet, then place them into KSOM in a culture dish after three washes in KSOM droplets. The transfer pipet's connecting tubing and filter can be controlled by mouth, a 1-mL syringe, or simply a rubber bulb.

Pronuclear transfer

Pronuclei exchanges between zygotes and reconstitution are achieved by nuclear transfer and electrofusion.

1. Zygotes are incubated for 20 min in the micromanipulation medium, HKSOM supplemented with cytochalasin D (2 μg/mL) and nocodazole (2 μg/mL), to render the membrane sufficiently elastic to withstand possible damage by an enucleation micropipet prior to micromanipulation. Enucleation and insertion of pronuclei (karyoplast) are performed on an inverted microscope with Nomarski optics and Narishige micromanipulators.

2. A few microdrops of micromanipulation medium are made on the micromanipulation chamber, covered with mineral oil. The chamber is placed on the microscope stage, and the microscope is focused on the edge of medium droplet. First the holding pipet and then the nuclear transfer pipet is mounted into the pipet holder at each side of the micromanipulator. Oil or Fluorinet is pushed to the tips of the pipet. The pipets are lowered in the medium droplets, in the middle field focused under the microscope at lower power (e.g., ×4).
3. A batch of zygotes is washed in the micromanipulation medium before being transferred to the droplets in the upper part of the chamber. At higher power (e.g., ×40), the holding pipet and nuclear transfer pipet are moved on the same focus plane as zygotes. A zygote is firmly aspirated by holding pipet using negative pressure and slightly turning back the microsyringe, ensuring appropriate orientation of pronuclei so that they are readily seen.
4. The spike of the enucleation pipette is gently inserted into the perivitelline space under the zona pellucida of the zygote without breaking the plasma membrane.
5. The beveled pipet tip is advanced adjacent to a pronucleus and then to another pronucleus. Male and female pronuclei are distinguishable based on their location and sizes. Female pronucleus is typically smaller and located closer to the second polar body in a vast majority of cases. By backward turning of a microsyringe, the pronuclei and the overlying membrane are drawn with a minimal amount surrounding cytoplasm into the pipette via gentle suction.
6. When the two pronuclei are inside the pipet, the pipet is slowly withdrawn through the zona pellucida, and the plasma membrane associated with pronuclei (karyoplasts) is pinched off from cytoplasm of enucleating zygotes with self-sealing. The enucleated zygotes are moved to the lower part of the chamber, and are released from the holding pipet by a little positive pressure from the syringe. To pick up an egg in the upper part and release it in the lower part, it is more convenient and efficient to move the microscope stage rather than the pipets.
7. The enucleation process is repeated for all zygotes, and the pronuclei can be released into a separate drop. After a batch of 15–20 zygotes are enucleated, a group of pronuclei-kayroplasts with minimal cytoplasm are created.

8. Using the enucleation pipet, the karyoplasts are then placed under zona pellucida through the same hole made during enucleation, and pushed into close contact with the cytoplasm of previously enucleated zygotes.
9. Similarly to GV nuclear transfer, pronuclear (PN) transfer also can be achieved by cutting the zona pellucida.
10. Micromanipulated embryos are washed several times in normal KSOM without inhibitors and then put back in the CO_2 incubator for recovery until fusion.
11. Electrofusion of karyoplasts and cytoplasts. Pronuclei karyoplasts coupled with enucleated cytoplasts (couplets) are equilibrated briefly (2–5 min) by washing in the fusion medium. The couplets are placed between two platinum electrodes (200 μm in diameter) by a distance of 200 μm, mounted in a fusion chamber, and aligned manually to make contact in a couplet parallel to the electrodes. Following an AC pulse (5V) for 2–3 s, a single DC pulse of 1.0–1.6 kV/cm for 20–60 μs is applied to induce fusion of the couplets. AC pulses enable contact between the plasma membrane of couplets and better alignment with electrodes, and thus increase fusion efficiency. Electric pulses are triggered using a BTX Electro Cell Manipulator. One hour later, success of fusion is achieved. If some couplets are not fused, they are subjected to another DC pulse for improving fusion efficiency. Efficiency of greater than 90% is routinely achieved by the electrical fusion.
12. Reconstituted zygotes are cultured in KSOM for 3 d. Embryos are checked for cleavage at 24 h, development to morula at 48 h, and development to blastocyst at 72–80 h.

Notes

1. This IBMX concentration prevents the GVBD without affecting oocyte viability and progression to metaphase II. Milrinone, a phosphodiesterase (PDE) inhibitor, can also be used in place of IBMX. Cytochalasin D is used to increase the oolemmal elasticity prior to micromanipulation.
2. There are many different designs of micromanipulators and inverted microscopes, some with motorized controls. Listed are what we used, although good alternative equipment is available, such as Eppendorf micromanipulators and inverted microscopes from Nikon, Leica, and Olympus. For the type of micro-manipulation described in this work, manually operated equipment is sufficient for easy manipulation and sensitivity of movement. In addition, the cost is

lower. Some old models listed may no longer be manufactured, but in principle, newer models should work similarly or better.

3. If vibration occurs and causes problems during micromanipulation, an antivibration table or air table is needed.
4. The quality of the microtools affects efficiency of micromanipulation. The outside diameter of holding pipet should be similar to or smaller than the size of oocytes or embryos including the zona pellucida, so that oocytes or embryos can make contact with the bottom of micromanipulation chamber during holding and manipulation. An appropriately sized pipett shaft can also provide better suction and produce better optical imaging of oocytes and embryos. The aperture of the holding pipet should be of appropriate size, generally approx 20% of the diameter of oocytes or embryos, to be able to immobilize them.
5. If the tip is not clean, i.e., contaminated with glass dust, it should be cleaned with hydrofluoric acid. The pipet is connected by soft tubing with a clean syringe and beveled tip is rinsed for a few seconds with dilute (10%, v/v) hydrofluoric acid prepared in a plastic tube. The acid also thins the outside surface of the tip, while the air is expelled through the pipet. The tip is washed by rinsing with sterile distilled water many times by changing water bottles, and then is washed with 100% ethanol and dried.
6. The tip of the pipet also can be coated with Nonidet P-40 (NP-40), rinsed a few times through three changes of distilled water, and dried in a warm oven. If NP-40 is no longer commercially available, Sigma offers comparable liquid, designated as Igepal CA-630.
7. In vitro maturation of denuded oocytes requires pyruvate for energy supplementation. Without cumulus cells, oocytes do not require gonadotropin hormones for maturation. GVBD normally occurs within 2–3 h of maturation, and >90% of these extrude a polar body and progress to MII stage after 15–16 h of maturation culture. Oocytes are stripped of cumulus cells and fixed for immunocytochemistry. Microtubules are stained by anti-β tubulin and/or anti-α tubulin and with fluorescein isothiocyanate (FITC)-conjugated anti-mouse IgG, actin filaments with Texas Red-conjugated Phalloidin, and DNA with Hoechst 33342. The samples are mounted onto a slide under a coverslip in the Vectashield mounting medium, and observed using a Zeiss fluorescence microscope and images are captured by an AxioCam using AxioVision 3.0 software.

8. Oocytes are secured by suction of holding pipet using a syringe. A sharpened, beveled pipet is penetrated through zona pellucida, and approaches GV. The GV, surrounded by a small amount of cytoplasm (*karyoplast*), is removed by the beveled pipet. During enucleation, the smallest amount of surrounding cytoplasm is removed, in order to maximize the positive effect of the host cytoplasm on subsequent nuclear maturation. The advantage of the second method is that no grinder is needed for beveling the pipets, and the disadvantage is two steps are required for removal of a GV. The opposite is for the first method. Yet, both methods produce similar efficiency of micromanipulation, in terms of speed of micromanipulation and survival of manipulated oocytes.
9. An alternat've method of PN nuclear transfer is first to cut the zona pellucida with a glass needle and then transfer pronuclei using blunt nuclear transfer pipet. The glass needle is inserted into the perivitelline space of zygotes, moved underneath of the holding pipet, and rubbed up and down against the wall of the holding pipet to cut the zona. A group of zygotes are cut and open holes for their zona with very short time. A blunt pipet is used to transfer pronuclei by insertion through the cut opening of zona, and then the pronuclei are transferred as described for beveled pipets. This method does not need to make a beveled, sharpened pipet, and thus does not need a micro-grinder. This method also dose not need to clean the pipette using acid and washing with water. Yet, the efficiency of these two methods for pronuclear transfer is similar.
10. The authors recommend referring to both GV nuclear transfer and PN transfer for similarities and differences in micromanipulation. In GV nuclear transfer, the nuclear transfer pipet is slightly large than for PN transfer, because of the relatively large size of GV.

10

Chromosomal Analysis

Telomeres are nucleoprotein structures that cap the ends of linear eukaryotic chromosomes. Telomeres are composed of DNA tandem repeats of a G-rich sequence, TTAGGG, in vertebrates, and are characterized by ending in a 3' G-rich overhang, which can fold back and invade the double-stranded region of the telomere, forming the so called T-loop. Because telomeres protect the chromosome ends from being recognized as DNA damage, and from being processed as such, excessive telomere shortening or telomere dysfunction triggers a DNA damage response leading to cell arrest or apoptosis.

Telomeres and Telomerase

As a consequence of the end-replication problem, cultured normal human somatic cells undergo telomere shortening with each cell division, resulting in the progressive erosion of telomeric repeats. Telomerase is an enzyme that is able to synthesize telomere repeats *de novo*, thus compensating for the end-replication problem in those cells in which it is expressed. Telomerase is absent from most somatic tissues but present in more than 90% of human tumors. Tissue homeostasis in long-lived species is maintained by cell division. Thus, in the absence of telomerase, progressive telomere shortening in somatic tissues may lead to the accumulation of cells with dysfunctional telomeres, eventually compromising tissue function. Highly proliferative tissues undergo telomere shortening as they age despite the expression of moderate levels of telomerase activity. In contrast, telomere shortening is virtually absent in less proliferative tissues such as skeletal muscle or in tissues with high telomerase activity such as male germ cells. Critically short telomeres correlate with the onset of age-related disease, such as vascular dementia, atherosclerosis, and cancer. The telomerase-deficient mouse model has been instrumental in dermining the biological consequences of telomere shortening. Telomerase-deficient mice (Terc$^{-/-}$) were obtained by elimination of the gene encoding murine Terc. Because telomerase activity is absent from most human adult somatic tissues, Terc$^{-/-}$ mice with short telomeres have been a useful model of telomere-driven aging in humans. Terc$^{-/-}$ mice show progressive telomere shortening with increasing generations, and only a limited number of generations can be obtained because of telomere exhaustion and increased chromosome end-to-end fusions. The pathologies associated with telomere dysfunction in Terc$^{-/-}$ mice affect most of the mouse tissues, such as the germ line, gut, skin, immune system, bone marrow, liver, heart, and blood vessels. Disease states in these mice are characterized by decreased cell proliferation and/or increased apoptosis. These observations support the notion that telomere length is a mitotic clock that limits replicative capacity of somatic tissues. In this context, telomere length may be an important marker of aging and age-related pathologies.

The most extended method of telomere length analysis is the *telomere restriction fragment* (TRF) analysis or telomeric Southern blot. However, this method has multiple drawbacks (discussed later), which led to the development of alternative methods for measuring telomere length based on fluorescence *in situ* hybridization (FISH). More recently, additional methods have been developed, such as hybridization protection assay, hybridization assay, and primed *in situ* (PRINS) hybridization, as well as PCR-based methods (STELA and quantitative-PCR). Here, we will focus on the application of TRF and FISH methods.

Telomere Restriction Fragment Analysis

The first approaches to measure telomere length were based on Southern blot hybridization using probes against telomere repeats. Telomere restriction fragment analysis, also known as telomeric Southern blot, relies on the differential digestion of genomic DNA using restriction enzymes, none of which can digest telomeric repeats. Hence, digestion of genomic DNA with a frequently cutting restriction enzyme will degrade genomic DNA to short fragments, leaving the bulk of telomeres intact—the so-called telomere restriction fragment. Following separation of genomic fragments by gel electrophoresis, and the identification of telomere-containing fragments by Southern blot with a telomere probe, the average size of TRFs reflects the mean telomere length of the sample. In order to apply this technique to samples from organisms with long telomeres, such as inbred laboratory mice, it is necessary to embed cells in agarose plugs for digestions with protease and restriction enzymes, and to use pulse field gel electrophoresis to resolve the TRFs.

TRF analysis has been the standard and most common method for telomere length analysis. The advantages of TRF analysis are: (i) it is a well known and widely used technique for telomere length studies, (ii) it has no special requirements in terms of reagents or equipment, (iii) it works with a wide variety of sample types including genomic DNA, cultured cells, and tissues (some of them require nuclei isolation), and (iv) it can measure a wide range of telomere lengths if combined with pulse-field agarose gel electrophoresis. In addition, this method can also be used to measure the G-strand overhang of telomeres. However, TRF analysis has numerous drawbacks: the technique (1) is time-consuming, difficult to quantify, and requires many cells (more than 10^5), (2) only yields an average value of telomere length of the whole sample, i.e., neither individual telomeres nor the mean telomere length values of individual cells are distinguished, (3) it is difficult to detect short telomeres, something that is crucial for aging studies, due to the electrophoretic migration features of short fragments, which are less concentrated and give a lower hybridization signal, (4) TRFs not only contain telomeres but also subtelomeric sequences, which can vary in length depending on the last restriction site at a given chromosome arm, thus increasing the heterogeneity of the TRFs and masking the "real" length of TTAGGG repeats.

Fluorescence In Situ Hybridization Methods

Telomere length analysis was considerably improved with the development of FISH methods. As compared to TRF analysis, FISH methods provide increased sensitivity, specificity, and resolution. This allows the measurement of telomere length at individual chromosomes at the single-cell level both in native tissue preparations and histological sections. Telomere length analysis by FISH is based on the specific labeling of telomeres with fluorescent *peptide nucleic acid* (PNA) oligonucleotide probes. PNA probes are synthetic peptides homologous to DNA, in which the negatively-charged phosphate-pentose backbone of the DNA is replaced by an uncharged N-2 amine ethyl-glycine backbone. This modification yields an extraordinarily stable, efficient, and specific hybridization of the probe to the target DNA. Because the intensity of the fluorescence signal from telomeric PNA probes that hybridize to a given telomere is directly related to telomere length, this method permits accurate telomere length measurements. FISH methods for telomere length measurements can be divided into two groups, namely quantitative (Q)-FISH, based on digital fluorescence microscopy, and Flow-FISH, based on flow cytometry.

Quantitative fluorescence *in situ* hybridization

The measurement of telomere length by Q-FISH is based on the use of digital fluorescence microscopy to determine telomere fluorescence after hybridization of metaphase spreads with a fluorescent PNA telomeric probe. By capturing individual telomere fluorescence signals, Q-FISH measures the lengths of all telomeres on metaphase spreads yielding a distribution of telomere length frequencies. Because metaphases are stained both for telomeres (e.g., with a telomeric PNA probe labeled with Cy3) and for chromosomes (e.g., with 4,6-diamidino-2-phenylindole dihydrochloride [DAPI]), analysis of the captured images yields information on telomere length distributions for each chromosome pair and chromosome arm. Thus, in contrast to TRF analysis, the Q-FISH technique permits measurement of telomere length at each individual chromosome end. The low detection limit of at least 0.15 kb further allows quantification of the number of "*signal-free ends*," i.e., chromosome ends with critically short telomeres (<0.15 kb). This is of particular relevance to aging studies since it has become increasingly clear that not the mean telomere length but rather the frequency of critically short telomeres, i.e. signal-free ends, is determinant for telomere dysfunction, loss of cell viability, and the ensuing onset of age-related pathologies. A further advantage of the Q-FISH method as compared to TRF analysis is its higher accuracy, resulting from the direct relation between the quantity of telomere repeats and the measured fluorescence signal. Furthermore, Q-FISH requires less starting material than TRF analysis. In general, 20 analyzed metaphases are enough to provide accurate telomere length measurements, whereas TRF analysis requires at least 10^5 cells.

Drawbacks of the Q-FISH technique are that (1) in its present form, it is a time-consuming and labor-intensive technique; (2) it requires a rather expensive and technically demanding fluorescence microscopy imaging system; (3) it yields as direct readout arbitrary units (integrated fluorescence intensity values) and, thus, to obtain absolute telomere length values, requires external calibration, using plasmids with cloned telomere repeats of defined length or cell lines, which stably maintain a defined and known telomere length distribution; and (4) because it relies on the preparation of metaphase spreads, it is restricted to actively proliferating cells in culture and cannot be applied to post-mitotic, differentiated, or senescent cells.

The development of Q-FISH protocols for cells in interphase circumvents the constraint of working with cells in culture and allows telomere length determinations in freshly isolated cells and in paraffin-embedded or cryostatic tissue sections as well. In both cases, Q-FISH can be combined with immunostaining techniques, allowing the simultaneous determination of telomere length and other parameters, such as markers of DNA damage or aging. A disadvantage of interphase Q-FISH measurements is that it is not possible to measure telomere signals on individual chromosomes, which precludes the quantification of signal free ends.

Flow fluorescence in situ hybridization

The Flow-FISH method for telomere quantification is based on the determination of telomere fluorescence in individual interphase cells using *fluorescence-activated cell sorting* (FACS) technology. In the Flow-FISH technique, cell suspensions are hybridized with a telomeric fluorescein isothiocyanate (FITC)-conjugated PNA probe and counterstained for DNA to normalize for DNA content. Sample acquisition by flow cytometry provides telomere length values in fluorescence units per counted cell. According to the cell cycle profile of the cell population, two signals for telomere fluorescence, one derived from cells in G2/M (4N DNA content) and one from cells in G0/G1 (2N DNA content), are obtained. The G0/G1 cell population is used for normalization to compensate for variations in the DNA content between different samples. As with the Q-FISH technique, to obtain absolute telomere length values, external calibration using cell lines, which stably maintain a defined and known telomere length distribution, is required.

Flow-FISH is a powerful method for large-scale telomere length analysis with potential clinical applications because (1) FACS technology is simple and amenable to automatization, providing all the features of a high-throughput technique, (2 the method is quantitative, reproducible, and accurate, and (3) Flow-FISH analysis of telomeres can be combined with the detection of cell surface markers. Limitations of the Flow-FISH method are that (1) its application is restricted to isolated cells, (2) most available Flow-FISH protocols focus on the analysis of hematopoietic cells, (3) the technique requires specialized and expensive equipment (flow cytometers and sorters), (4) in contrast to Q-FISH, which provides information on the presence of short telomeres, Flow-FISH measurements only provide a mean value for the telomere length of the whole population of telomeres within a cell, and (5) the presence of intra-chromosomal telomeric signals may cause an overestimation of telomere length.

MATERIALS

TRF Analysis

Reagents

1. Phosphate-buffered saline (PBS; 137 m*M* NaCl, 2.7 m*M* KCl, 4.3 m*M* Na_2HPO_4, 1.4 m*M* KH_2PO_4, pH 7.3).
2. 1.5 mL Eppendorf tubes.
3. 2% (w/v) low melting preparative grade agarose in PBS.
4. Proteinase K solution: 20 mg/mL proteinase K in a 10 m*M* Tris-HCl buffer (pH 8.0) containing 1 m*M* EDTA, 0.3 *M* sodium acetate, and 0.2% (w/v) sodium dodecyl sulfate (SDS). Aliquots of this stock solution are stored at –20°C. Proteinase K working solutions are prepared freshly by diluting the proteinase K stock in the same buffer 10-fold to 2 mg/mL.
5. TE 1X buffer: 10 m*M* Tris-HCl (pH 8.0) containing 1 m*M* EDTA.
6. 100 m*M* phenylmethylsulphonyl fluoride (PMSF) in isopropanol. Aliquots are stored at –20°C.
7. MboI restriction enzyme and MboI 10X buffer.
8. Pulse-field certified agarose.
9. TBE 1X buffer: 89 m*M* Tris base, 89 m*M* boric acid, and 2 m*M* EDTA in H_2O.
10. Molecular weight DNA markers: ProMega-Markers Lambda ladders, DNA size standards-5 kb ladder, 1 kb DNA ladder.
11. Low melting SeaPlaque agarose.
12. Ethidium bromide stock at 10 mg/mL.
13. Denaturation solution: 1.5 *M* NaCl and 0.5 M NaOH in H_2O.
14. Neutralization solution: 1.5 *M* NaCl in a 0.5 *M* Tris-HCl buffer (pH 7.5).
15. SSC 2X: 0.3 *M* NaCl in a 0.03 *M* sodium citrate buffer (pH 7.0).
16. Hybond-N^+ nylon membrane.
17. HIGH buffer: 1% (w/v) bovine serum albumin (BSA), 0.2 *M* sodium phosphate, 15% (w/v) formamide, 1 m*M* EDTA, and 7% (w/v) SDS in H_2O.

18. ^{32}P-labeled telomeric probe: a random-primed-labeled 1.6-Kb fragment of the $(TTAGGG)_n$ sequence, obtained by restriction of pNYH3 1.6 plasmid, or, alternatively, T4-polynucleotide kinase-labeled $(TTAGGG)_4$ or $(CCCTAA)_4$ oligonucleotides.
19. High-stringency wash solution: 0.2X SSC (diluted from reagent 15) supplemented with 0.1% (w/v) SDS.

    The reagents listed above are required for TRF analysis of cells in culture or of isolated cells that have been obtained by disaggregation of soft tissues. When tissues are used that cannot be disaggregated into isolated cells, in addition the reagents listed below are required to extract nuclei.
20. Solution 1: 15 m*M* Tris-HCl (pH 7.5) supplemented with 60 m*M* KCl, 15 m*M* NaCl, 0.5 m*M* spermine, 0.15 m*M* spermidine, 14 m*M* β-mercaptoethanol, 2 m*M* EDTA, and 0.3 *M* sucrose (store at 4°C).
21. Solution 2: 15 m*M* Tris-HCl (pH 7.5) supplemented with 60 m*M* KCl, 15 m*M* NaCl, 0.5 m*M* spermine, 0.15 m*M* spermidine, 14 m*M* β-mercaptoethanol, 1 m*M* EDTA, and 1.37 *M* sucrose (store at 4°C).
22. Solution 3: 15 m*M* Tris-HCl (pH 7.5) supplemented with 60 m*M* KCl, 15 m*M* NaCl, 0.5 m*M* spermine, 0.15 m*M* spermidine, 14 m*M* β-mercaptoethanol, and 1 m*M* EDTA (store at 4°C).
23. 2% (w/v) preparative grade low-melting agarose in solution 3.

Equipment and supplies

1. Bio Rad CHEF disposable plug mold.
2. CHEF-DR II pulsed-field electrophoresis system.
3. Standard gel casting stand.
4. Gel-doc 2000 system with Quantity-One software.
5. UVC 500 ultraviolet (UV) crosslinker.
6. Film suitable for ^{32}P-autoradiography, such as Kodak X-OMAT. When, instead of conventional autoradiography, phosphor screens are used to detect ^{32}P-labeled TRFs, a Storm 820 Phosphorimager with phosphor screens, and ImageQuant software is required.

   The equipment listed above is required for TRF analysis of cells in culture or of isolated cells that have been obtained by disaggregation of soft tissues. When tissues are used that cannot be disaggregated into isolated cells, in addition the equipment listed below is needed.
7. Tissue homogenizer Ultra-Turrax T8.

8. Refrigerated high-speed centrifuge with swing-out rotor and 10-mL tubes.
9. Sterile gauze and syringe.

Q-FISH Analysis

Reagents

1. Colcemid Karyomax 10 μg/mL.
2. Phosphate-buffered saline (PBS).
3. Trypsin-EDTA: 0.05% Trypsin, 0.53 m*M* EDTA.
4. Hypotonic solution: 0.03 *M* sodium citrate in H_2O, prewarmed to 37°C.
5. Methanol-acetic acid fixative solution: freshly prepared mixture of 3 volumes of methanol and 1 volume of glacial acetic acid.
6. 100% methanol.
7. 4% (v/v) formaldehyde in PBS.
8. Acidified pepsin solution: 1 mg/mL pepsin. In 10 m*M* HCl, prewarmed to 37°C.
9. Ethanol (absolute, 100%).
10. 90% (v/v) ethanol in H_2O.
11. 70% (v/v) ethanol in H_2O.
12. Deionized formamide: continuously stir for 1 to 2 h 10% (w/v) AG 501 X8 resin (20–50 mesh) in formamide (ultra-pure from Fluka). The resin is removed by filtration, and aliquots are stored at –20°C. Avoid repeated freeze-thawing.
13. Cy3-labeled $(CCCTAA)_3$ PNA telomeric probe: prepare lyophilized aliquots of 5 μg. Store at –20°C. Prior to use, reconstitute with 200 μL ddH_2O to obtain a 25 μg/mL stock solution. Store at 4°C.
14. Hybridization solution: 10 m*M* Tris-HCl (in ddH_2O, pH 7.0) containing 70% (v/v) deionized formamide, 0.25% (w/v) blocking reagent for nucleic acid hybridization and detection, 5% (v/v) of $MgCl_2$ buffer solution (25 m*M* $MgCl_2$, 9 m*M* citric acid, 82 m*M* Na_2HPO_4), and 0.3 μg/mL Cy3-PNA telomeric probe.
15. Wash solution 1: 70% (v/v) formamide in 10 m*M* Tris-HCl (pH 7.2), supplemented with 0.1% (w/v) BSA.
16. Wash solution 2: 0.1 *M* Tris-HCl (pH 7.2) containing 0.15 *M* NaCl, and 0.08% (v/v) Tween 20.
17. DAPI solution: mix 2-3 volumes of "Vectashield mounting media for fluorescence" (w/o DAPI) with one volume of "Vectashield mounting media for fluorescence with DAPI".

18. Nail polish.
19. Calibration beads: latex Fluospheres, carboxylate-modified, 0.2 μm diameter.
20. Slow Fade Antifade Kit.
21. Xylene (only needed when paraffin sections are processed).

Equipment and supplies

1. Standard cell culture equipment and supplies.
2. Water bath at 37°C.
3. Platform mixer.
4. Glass pasteur pipets.
5. Slides and coverslips.
6. Coplin jars.
7. Heating plaque.
8. FISH image acquisition system.

Flow-FISH Analysis

Reagents

1. PBS without Ca^{2+} and Mg^{2+}.
2. Deionized formamide: continuously stir for 1 to 2 h 10% (w/v) AG 501 X8 resin (20-50 mesh) in formamide. Aliquots are stored at –20°C. Avoid repeated freeze-thawing.
3. FITC-labeled $(CCCTAA)_3$ PNA telomeric probe (Applied Biosystems, Foster City, CA): prepare lyophilized aliquots of 10 μg and store at –20°C. Prior to use, reconstitute with 100 μL ddH_2O to obtain a 100 μg/mL stock solution. Store at 4°C.
4. Hybridization buffer: 20 m*M* Tris-HCl (pH 7.2), containing 70% (v/v) deionized formamide, 1% (w/v) BSA, and 0.3 μg/mL FITC-PNA telomeric probe. For negative controls, hybridization buffer without addition of the FITC-PNA telomeric probe is used.
5. Wash solution 1: 20 m*M*Tris-HCl (pH 7.2), containing 70% (v/v) deionized formamide, 0.1 % (w/v) BSA, and 0.1 % (v/v) Tween 20.
6. Wash solution 2: PBS supplemented with 0.1 % (w/v) BSA and 0.1 % (v/v) Tween-20.
7. PI solution: PBS supplemented with 0.1% BSA, 10 μg/mL RNAse, and 0.1 μg/mL propidium iodide.

Equipment and supplies

1. Flow cytometer and supplies for flow cytometry.
2. Thermomixer.

METHODS

TRF Analysis

Sample preparation

Protocol for cells in culture or isolated cells obtained by disaggregation of soft tissues

1. Freshly harvested cells (1.5×10^6) are resuspended in PBS, pelleted in a tabletop centrifuge at 2500*g* at room temperature, washed twice with PBS, and pelleted again. Cell pellets are snap-frozen in liquid nitrogen and stored at −80°C until further use.
2. Just prior to the preparation of agarose plugs (discussed later), cell pellets are resuspended in 50 μL of PBS and incubated for 5 min at 50°C.
3. Add 50 μL of a prewarmed (50°C) solution of Bio-Rad preparative grade low melting agarose (2% [w/v] in PBS).
4. Mix well and incubate for further 5 min at 50°C.
5. Continue with the preparation of agarose plugs as discussed later.

Protocol for tissues that cannot be disaggregated into isolated cells

1. Freshly isolated tissue samples are snap-frozen in liquid nitrogen and stored at −80°C until used.
2. A piece of frozen tissue (sufficient to isolate 2×10^6 nuclei) is resuspended in 3 mL of solution 1 using a tissue homogenizer.
3. Filter tissue homogenate through three to four layers of sterile gauze packed into a 5-mL syringe.
4. Layer filtered homogenate over 5 mL of solution 2.
5. Centrifuge in a swing-out rotor for 15 min at 20,000*g* at 4°C.
6. Resuspend the pellet of nuclei in 1 mL of solution 1, count nuclei, and remove an aliquot of 1.5×10^6 nuclei for further analysis.
7. Pellet nuclei by centrifugation at 5000*g* for 10 min in a tabletop centrifuge.
8. Resuspend pelleted nuclei in 50 μL of solution 3 and incubate at 50°C for 5 min.
9. Add 50 μL of a prewarmed (50°C) solution of Bio-Rad preparative grade low melting agarose (2% [w/v] in solution 3).
10. Mix well and incubate for 5 min at 50°C.

Preparation of agarose plugs

1. Transfer the prewarmed (50°C) agarose solution containing either cells or isolated nuclei to the Bio-Rad CHEF disposable plug

molds, allow to cool down for 5 min at room temperature, and for further 15 min at 4°C.

2. Transfer the plugs to 1.5-mL Eppendorf microcentrifuge tubes.
3. Add 500 μL of proteinase K working solution (2 mg/mL) and incubate overnight at 50°C.
4. Wash plugs two times for 1 h with 1X TE and, subsequently, once for 1 h with 1X TE containing PMSF (1 m*M*) to inactivate proteinase K and, finally, once for 1 h with 1X TE.
5. Plugs can now be stored at 4°C in 1X TE until used.

MboI restriction digestions

1. Wash the agarose plugs for 1 h in 1 mL H_2O on a shaking platform.
2. Equilibrate the plugs for 1 h in 1 mL of 1X MboI restriction buffer on a shaking platform.
3. Incubate the plugs overnight at 37°C in 300 μL of 1X MboI buffer in the presence of 50 U MboI.

Pulse-field electrophoresis

1. Wash agarose plugs for 30 min in 500 μL of distilled water on a shaking platform.
2. Equilibrate the plugs for 60 min in 0.5 mL of 0.5X TBE.
3. Prepare a 1% agarose gel in 250 mL 0.5X TBE using the Bio-Rad standard gel casting stand.
4. Prior to electrophoresis, circulate approx 3 L of 0.5X TBE for 30 min through the pulse-field electrophoresis tank.
5. Cut the plugs into three equal pieces, and load one of them onto the gel. The remaining pieces of the gel plug can be stored in 1X TE at 4°C for further electrophoresis runs. Load the agarose-embedded type molecular weight DNA markers, and cover all slots, except the ones that will be used for the liquid type molecular weight DNA markers, with low-melting SeaPlaque agarose (1% in 0.5X TBE). Allow agarose to solidify.
6. Mount the caster with the gel in the pulse field chamber and load the liquid type molecular weight DNA markers.
7. Run the gel without recirculation for 1 h until the liquid type molecular weight DNA marker has entered into the gel, and then set the system to the following parameters: initial pulse, 5 s; final pulse, 5 s; electrical current, 6 V/cm^2; run time, 23 h (for mouse cells) or 18 h (for human cells).

8. After 1 h, restart the recirculation of the electrophoresis buffer setting the cooling module to 14°C. Continue electrophoresis for the remaining time.
9. Following electrophoresis, remove the gel and stain it under gentle agitation for 20 min at room temperature in 1 L of 0.5X TBE supplemented with 50 μL of ethidium bromide stock solution (10 mg/mL).
10. Wash the gel briefly in 0.5X TBE, and photograph it under UV light side-by-side with a ruler (needed later for quantification of the telomere size).
11. Wash the gel under gentle agitation three times for 30 min in denaturation solution.
12. Wash the gel under gentle agitation three times for 30 min in neutralization solution.
13. Equilibrate the gel for 5 min in 2X SSC.

Southern blot

1. Wet-transfer the gel in 2X SSC overnight to a Hybond-N^+ nylon membrane.
2. UV-crosslink the DNA to the membrane at 120 mJ/cm^2.
3. Prehybridize the membrane with HIGH buffer for 1 h at 65°C.
4. Add the $(TTAGGG)_n$ telomeric probe to the pre-hybridization solution and hybridize overnight at 65°C. When $(TTAGGG)_4$ or $(CCCTAA)_4$ oligonucleotide probes are used, perform pre-hybridization and hybridization.
5. Wash the membrane three times for 30 min with prewarmed (65°C) high- stringency wash solution.
6. Autoradigraph the membrane using a Kodak X-OMAT film or, alternatively, expose the membrane to phosphor screens for 30 min to overnight, and visualize on a Storm 820 PhosphorImager using ImageQuant software.

Data analysis

The average TRF length of each sample can be calculated both by densitometric analysis of autoradiographs or PhosphorImager images, using the Quantity-One and ImageQuant software package provided with the GelDoc 2000 system and Storm 820 PhosphorImager, respectively. Please note that, for detailed analysis of TRF signal intensity taking into account subtelomeric contributions, a Microsoft Exce spreadsheet called "Telorun" is available at the homepage of the Shay and Wright laboratory.

The general principle of TRF analysis using MboI digestion of genomic DNA from primary murine embryonic fibroblasts followed by pulse field electrophoresis and Southern blotting. Hybridization of digested genomic DNA with a ^{32}P-labeled telomeric probe yields a smear that corresponds to the distribution of the entire population of telomere lengths within the pool of analyzed cells.

Densitometry and comparison with the position of molecular weight markers yields an estimate of the mean telomere length (about 30 kb) and provides information on the presence of short telomeres (<9 kb). Germline re-introduction of telomerase in late-generation telomerase-deficient (Terc$^{-/-}$) mice via crosses with Terc$^{+/-}$ mice affects telomere length in the progeny. TRF analysis shows low-molecular-weight TRFs (<9 kb) in the Terc$^{-/-}$ sire that are absent in the Terc$^{+/-}$ dam. Littermates that received the Terc-null allele from the dam (Terc$^{-/-}$ progeny) showed increased telomere shortening as evidenced by the presence of TRFs below 7 kb (indicated with asterisks), whereas littermates that received the wild-type allele (Terc$^{+/-}$ progeny) had longer telomeres with low molecular weight TRFs that were clearly above 7 kb.

Q-FISH Analysis

Q-FISH analysis of metaphase spreads

Metaphase Preparation

1. For metaphase preparations, plate an appropriate number of cells so that they will reach subconfluence with the maximal number of dividing cells by the start of sample preparation. To subconfluent and actively dividing cells in culture, add colcemid at a final concentration of 0.1 μg/mL to arrest cells in metaphase.
2. Harvest cells in suspension by gentle centrifugation (400*g*). Harvest adherent cells by trypsinization.
3. Pellet cells by gentle centrifugation (400*g*).
4. Wash cells with PBS.
5. Pellet cells by gentle centrifugation (400*g*), and remove the supernatant, leaving 1 mL to resuspend the cell pellet.
6. With a glass pasteur pipet, under continuous and gentle mixing, add dropwise 9 mL of prewarmed (37°C) hypotonic solution.
7. Incubate in a water bath at 37°C for 25 min.
8. Pellet cells by gentle centrifugation (400*g*), and remove the supernatant, leaving 1 mL to resuspend the cell pellet.

9. With a glass pasteur pipet, under continuous and gentle mixing, add dropwise 9 mL of freshly prepared methanol-acetic acid fixative solution.
10. Repeat steps 8 and 9 twice (pelleting and fixation of cells). Fixed cells can be stored at –20°C indefinitely until further use. Prior to use, stored cell have to be resuspended in freshly prepared methanol-acetic acid fixative solution.
11. Pellet cells by gentle centrifugation (400*g*), and remove the supernatant, leaving enough fixation solution to allow resuspension of cells at a final concentration of 1-5 × 10^6 cells/mL.
12. Prepare slides by washing them gently with 100% methanol.
13. Prepare thin glass capillaries from pasteur pipets by heating them over a Bunsen flame.
14. Dispose 1 mL of 45% acetic acid on the slide surface and drip off the acetic acid solution.
15. Using a thin glass capillary (from step 13), immediately spot small drops of fixed cells all over the slide.
16. Allow excess of fixative solution to drain off and air-dry slide overnight.
17. Wash and rehydrate slides in a coplin jar with PBS for 15 min at room temperature under gentle shaking.
18. Fix slides in a coplin jar with formaldehyde (4% [v/v]) in PBS) for 2 min at room temperature under gentle shaking.
19. Wash slides in a coplin jar three times with PBS for 5 min at room temperature under gentle shaking.
20. Perform digestion treatment of slides in a coplin jar with prewarmed (37°C) acidified pepsin solution for 10 min under gentle shaking in a water bath at 37°C.
21. Wash slides in a coplin jar twice with PBS for 5 min at room temperature under gentle shaking.
22. Fix slides in a coplin jar with formaldehyde (4% [v/v] in PBS) for 2 min at room temperature under gentle shaking.
23. Wash slides in a coplin jar three times with PBS for 5 min at room temperature under gentle shaking.
24. Dehydrate samples by subsequent treatment with 70%, 90%, and 100% ethanol in coplin jars, each for 5 min at room temperature under gentle shaking.
25. Allow slides to air-dry for at least 20 min at room temperature.

26. Place two 12.5-μL drops of hybridization solution, which contains the Cy3- PNA telomeric probe, on a coverslip and, avoiding air bubbles, place the dry slide upside down on the coverslip.
27. Denature samples on a heating plaque at 80°C for 3 min.
28. Place slides in a plastic slide box, put the box into a parafilm-sealed wet chamber, and incubate in the dark for 2 h at room temperature.
29. Wash slides twice, in a coplin jar with wash solution 1 for 15 min in the dark at room temperature under continuous shaking. Do not remove the coverslips. After 10 min in the first washing step, they fall away from the slide upon lifting the slide baskets.
30. Wash slides three times in a coplin jar with wash solution 2 for 5 min in the dark at room temperature under continuous shaking.
31. Dehydrate samples by subsequent treatment with 70%, 90%, and 100% ethanol in coplin jars, each for 5 min in the dark at room temperature under continuous shaking.
32. Allow slides to air-dry for at least 20 min at room temperature, avoiding exposure to light.
33. Prepare DAPI solution and place two 12.5-μL drops of hybridization solution on to a coverslip and, avoiding air bubbles, place the dry slide upside down on the cover slip. Allow to sit for some minutes until the DAPI solution has spread all over the sample.
34. Seal coverslips with nail polish and allow to air-dry for some minutes.
35. Sealed slides can be stored in the dark at 4°C for up to 3 d until examination by digital fluorescence microscopy.

Image acquisition and data analysis

For Q-FISH analysis, images of telomeres and chromosomes can be acquired with any standard FISH image acquisition system that fulfills the criteria described in detail previously. In our laboratory, we use a Leica DM RA2 fluorescence microscope equipped with a PL Apo 100×/1.40-0.7 oil iris 0.09 ∞/0.17 objective, excitation and emission filters for Cy3 (Leica filter cube Y3: excitation filter BP 545/30 nm, dichromatic mirror 565 nm, suppression filter BP 6 10/75 nm) and DAPI (Leica Filter cube A: excitation filter BP 340-380 nm, dichromatic mirror 400 nm, suppression filter LP 425 nm), and HBO 100W/2 mercury lamps. Image acquisition is performed with a *charge-coupled device* (CCD) camera at 700 × 500 pixels resolution and 8-bit depth. Images are processed with the Leica Q-FISH software and analyzed with the TFL-TELO software, a gift of Drs. Lansdorp and Poon.

To determine the optimal exposure range to capture telomeric signals, metaphases from L5 178Y-S and L5 178-R murine leukemia cell lines, that present stable and known telomere lengths of 10 and 79 kb, respectively, are used. Metaphases from these cells were processed by Q-FISH and captured at different exposure times (usually in the range from 0.05 to 1.0 s) until conditions are obtained in which short telomeres from L5178Y-S cells can be detected without reaching saturation of the strong telomere signal in L5178-R cells, i.e., there is a linear relation between telomere length and fluorescence intensity. In addition, to compensate for effects of variations in the signal intensity and homogenei.y of the light field over time, periodically capturing images of 0.2 μm carboxylate-modified latex Fluospheres is recommended. Images are captured using the Cy3-specific filter cube (Y3), and are analyzed with the "*spotIOD*" application of the TFL-TELO software.

To calibrate the system for telomere length measurements, metaphases from L5 178Y-S and L5 178-R cell lines are captured. The stable and known mean telomere length of these cells allows the conversion of arbitrary fluorescence units into DNA length units by linear regression. The slope of the linear regression of arbitrary units of fluorescence (auf) vs the length of telomeres in L5 178Y-R and L5 178Y-S cells is used to calculate telomere length of samples (Telomere length of a sample in kb = auf of the sample/slope of the L5 178Y-S/R regression curve).

For Q-FISH analysis, about 20 to 50 metaphases of each sample from at least two slides per sample are acquired and analyzed with the "*chromosome/telomere*" application of the TFL-TELO software. The obtained output files, which contain telomere length values sorted by chromosome pair and chromosome, are now integrated and analyzed with the Excel software to obtain telomere length distributions. *Note:* to be able to compare telomere lengths of different samples, these samples must be captured in the same session. To compare values obtained from different sessions, a linear regression with mean auf values for beads and L5178Y-S/L5178-R cells from the different sessions must be made. This implies that, in each session of image capturing, the following control samples and procedures must be included: (1) capture images of fluospheres to validate even illumination of the microscope field, (2) capture images of fluospheres and L5 178Y-S/L5

178-R cells to obtain reference values for eventual variations of signal intensity during each session, and (3) use the signal intensities of L5 178Y-S/L5 178-R cells to convert fluorescence intensities into telomere length (kb) values.

Q-FISH analysis of isolated interphase cells

When cells are not able to divide in culture or cannot be cultured for other reasons, Q-FISH analysis can be performed on interphase nuclei, described as follows.

Sample preparation

Freshly isolated cells (1-5 × 10^6 cells) are suspended in cell culture medium, pelleted, washed once with medium, and resuspended in a final volume of 1 mL of medium. Subsequently, cells are subjected to treatment with hypotonic solution, i.e., standard protocol for metaphase preparations described previously, and taken through the subsequent steps of this protocol.

This technique can also be used for interphase cells grown in slide chambers. In this case, the protocol is essentially the same as for Q-FISH with metaphase cells, but with the difference that cells keep attached to the slide and do not require colcemid and hypotonic treatment:

1. Remove cell culture media from chamber slide, and wash with PBS three times for 5 min at room temperature.
2. Remove PBS, leaving sufficient volume to cover the chamber slide surface.
3. Fix cells by adding slowly, and under constant and gentle mixing, freshly prepared methanol-acetic acid fixative solution.
4. Incubate for 5 min at room temperature and repeat step 3 twice.
5. Remove the fixative solution and allow slides to air-dry overnight.
6. Remove the chamber slide walls and proceed with the standard protocol for metaphase preparations described above, starting with step 18.

Image acquisition and data analysis

Q-FISH analysis of interphase nuclei is complicated by the fact that, in interphase, telomeres are distributed throughout the nucleus in different planes. Thus, it is essential to adjust the focus of the microscope so that most telomeres are in focus. With this strategy, about 100 nuclei are captured, and the obtained telomere images are analyzed with the "*spot IOD*" application of TFL-TELO. Prior to data integration in Excel, it is recommended that spikes and doublets in the signal histogram be removed.

Q-FISH analysis of paraffin and cryosections

Sample preparation

The adaptation of the standard Q-FISH protocol described previously for paraffin sections includes de-paraffination of samples with Xylene and subsequent rehydration:

1. Wash slides three times in a coplin jar with Xylene for 3 min at room temperature under gentle shaking.
2. Rehydrate samples by subsequent treatment of slides with 100%, 90%, and 70% ethanol in coplin jars, each for 3 min at room temperature under gentle shaking.
3. Wash slides in a coplin jar 3 times with PBS for 5 min at room temperature under gentle shaking.
4. Proceed with the standard protocol for metaphase preparations described above, starting with step 18.

The adaptation of the standard Q-FISH protocol for cryosections does not involve additional steps. The sections are simply washed in a coplin jar 3 times with PBS for 5 min at room temperature under gentle shaking.

Image acquisition and data analysis

Q-FISH analysis of nuclei in tissue sections is performed essentially as described previously for interphase nuclei.

Combined immunostaining and Q-FISH

Q-FISH analysis can be combined with immunostaining techniques and applied to both tissue sections and cultured cells. In general, prior to Q-FISH analysis, fixation and immunostaining of the samples are performed according to the recommendations of the manufacturer of the antibody. Subsequent to immunostaining, samples (e.g., tissue sections on slides or cells grown in slide chambers) are washed in a coplin jar three times with PBS for 5 min at room temperature under gentle shaking. Proceed with the standard protocol for metaphase preparations, but with the exception that pepsin treatment is omitted.

The following example for immunostaining Q-FISH analysis of tissue sections describes the simultaneous determination of γ-H2AX immuno-fluorescence and telomere length in cryosections of mouse testis:

1. Prepare testis cryosections and let air-dry overnight at room temperature.

2. Fix sections with cold acetone in a coplin jar for 2 min under gentle shaking at room temperature.
3. Permeabilize sections twice with PBS containing 0.1% (v/v) Triton-X100 in coplin jars for 15 min under gentle shaking at room temperature.
4. Block samples with 5% (w/v) BSA in PBS for 30 min at room temperature in a wet chamber.
5. Incubate with primary antibody (anti-phospho-histone H2AX [Ser139] mouse monoclonal IgG1 diluted 1:500 in a solution of 2% [w/v] BSA in PBS) for 1 h at room temperature in a wet chamber.
6. Wash three washes with PBS in coplin jars for 5 min under gentle shaking at room temperature.
7. Incubate with secondary antibody (Alexa Fluor488-labeled goat anti-mouse IgG [H+L], diluted 1:400 in a solution of 2% [w/v] BSA in PBS) for 1 h at room temperature in a wet chamber.
8. Wash three times with PBS in coplin jars for 5 min under gentle shaking at room temperature.
9. Proceed with the standard protocol for metaphase preparations described above, starting at step 18 and with the exception that pepsin treatment is omitted.

Q-FISH in combination with cytogenetic studies

Because telomere length is heterogeneous among chromosome pairs and arms, telomere dysfunction might involve only a subset of chromosomes. Thus, it may be desirable to quantify telomere length at specific chromosomes. In this particular case, the location of metaphases on a slide (*x*-*y* coordinates) is recorded and, subsequent to Q-FISH analysis, samples can be processed for Chromosome Painting or spectral karyotyping (SKY) analysis. Then it is possible to go back to each of the previously captured metaphases and to assign telomere length values to individual chromosome ends.

Both kits for chromosome painting (STAR*FISH chromosome painting kit) can be used subsequent to Q-FISH analysis. For this purpose, coverslips are removed by dissolving the nail polish with acetone, and slides are washed with PBS. Subsequently, samples can be processed following the manufacturer's protocols of the chromosome painting or SKY kit. Note that, when the SKY protocol is used subsequent to Q-FISH analysis, pepsin treatment is omitted from the SKY protocol.

Flow-FISH Analysis

Sample preparation

1. Pellet 10^6 cells in a 15-mL Falcon tube by centrifugation for 5 min at 400*g*.
2. Wash cells in PBS supplemented with 0.1% (w/v) BSA.
3. Pellet cells by centrifugation for 5 min at 400*g*.
4. Resuspend the pellet in 2 mL of PBS supplemented with 0.1% (w/v) BSA, and transfer sample to two 1.5-mL Eppendorf tubes.
5. Centrifuge for 15 s at 12,000*g*.
6. Resuspend pellets in 500 μL of hybridization buffer, one of them, which will serve as negative control, in hybridization buffer without the FITC-labeled telomeric PNA probe.
7. Denature samples for 10 min at 80°C under continuous shaking.
8. Incubate for 2 h in the dark at room temperature.
9. Centrifuge at 16°C for 8 min at 650*g*.
10. Wash samples twice with 1 mL of wash solution 1.
11. Centrifuge at 16°C for 8 min at 650*g*.
12. Wash once with 1 mL of wash solution 2.
13. Centrifuge at 16°C for 7 min at 300*g*.
14. Resuspend pellet in 500 μL of propidium iodide solution and incubate for 2 h at room temperature.
15. Analyze samples by FACS immediately or store samples at 4°C for a maximum of 2 d.

Flow cytometry

Flow-FISH analysis can be performed with any conventional flow cytometer. We have used both the FACSCalibur system with CellQuest software and the Coulter Flow Epics with software System2 as follows:

1. Cell cycle profiles are recorded selecting cells of adequate size (forward scatter) and complexity (side scatter), excluding dead cells and fragments from analysis. With the representation of area and width of the propidium iodide channel (FL2), select only single cells.
2. Through proper gating, only cells in G0/G1 (2N DNA content) are selected in the histogram of the propidium iodide channel.
3. The G0/G1 cells in the histogram for FL1, the channel for FITC, yield the distribution of whole-cell telomeric fluorescence intensity values, together with the descriptive statistics parameters. To compensate for the contribution of cellular auto-fluorescence,

fluorescence values for negative control cells, i.e., cells hybridized in the absence of the FITC-telomeric PNA probe, are subtracted.

4. As in the case of Q-FISH analysis, L5178Y-R y L5178Y-S cell lines are processed in parallel to the other samples. These cells serve as internal standard in order to compensate for variations in the system and to convert arbitrary units of fluorescence into telomere length values.

Notes

1. For human samples, instead of MboI, a mixture of the restriction enzymes HinfI and RsaI may be used to generate TRFs.
2. Instead of sodium citrate, especially in the case of human lymphocytes and bone marrow cells, KCl may be used for hypotonic treatment. Prepare a solution of 70 m*M* KCl in H_2O, and prewarm it to 37°C.
3. As an alternative to this Cy3-labeled PNA probe, an FITC-labeled $(CCCTAA)_3$ PNA telomeric probe may be used. However, it should be noted that the emission spectrum of FITC and DAPI, which is used for counterstaining of DNA, overlap to some degree (cross emission). Thus, in the case of working with FITC-labeled $(CCC\text{-}TAA)_3$, we strongly recommend using filter-cubes with bandpass filters (excitation filter: BP 480/40 nm, dichromatic mirror: 505 nm, suppression filter: BP 527/30 nm).
4. Instead of DAPI, which cannot be used together with blue fluorophores, propidium iodide may be used to stain chromosomes. Propidium iodide stock solutions are prepared by dissolving 100 μg/mL propidium iodide in water. Aliquots are stored at –20°C. Working solutions are freshly prepared by diluting the stock in antifade solution to a final concentration of 0.1 μg/mL. Note that propidium iodide cannot be used together with red fluorophores, such as Cy3- labeled telomeric probes. Thus, for telomere detection, propidium iodide staining of chromosomes has usually been combined with staining of telomeres with a FITC-labeled $(CCCTAA)_3$ PNA probe. Note that cross emission between propidium iodide and FITC is a major drawback of this method.
5. Alternatives to the mix of Vectashield mounting media recommended here are described in detail on the homepage of the Heslop-Harrison group.
6. More economic self-made alternatives to the Slow Fade Antifade Kit recommended here are described in detail on the homepage of the David Spector laboratory.

7. For TRF analysis of samples from organisms with long telomeres, such as inbred laboratory mice, it is essential to embed cells in agarose plugs for digestions with protease and restriction enzymes, in order to avoid fragmentation of genomic DNA, as it occurs during conventional DNA extraction procedures. Note that this procedure also considerably improves the quality of TRFs of human samples.
8. To simplify quantification of telomere size, one might wish to include radiolabeled molecular weight markers. A detailed protocol for the preparation of a ^{32}P-labeled high molecular weight DNA ladder is available at the homepage of the Stewart laboratory. Alternatively, a photo of the ethidium bromide stained gel with the DNA markers and a ruler that has been carefully aligned with the gel, is used to determine telomere length.
9. As an alternative to the classical Southern Blot-based TRF analysis described here, the slightly more sensitive in-gel hybridization method may be used. In this case, following ethidium bromide staining, gels are dried down on filter paper for 1 h at 50°C. Subsequently, gels are pre-hybridized for 1 h at 55°C in 20 m*M* NaH_2PO_4, 0.1% SDS, 5X Denhardt's reagent and 5X SSC. Gels are hybridized for 3 h to overnight at 55°C in 5 mL prehybridization solution with T4-polynucleotide kinase ^{32}P-labeled $(TTAGGG)_4$ or $(CCCTAA)_4$ oligonucleotide probes. Gels are washed three times for 20 min in 3X SSC at room temperature, and three times for 20 min in 3X SSC/0.1% SDS at 58°C. Gels are autoradiographed using a Kodak X-OMAT film or exposed to phosphor screens for 30 min to overnight.
10. The duration of the colcemid treatment depends on the cycling rate of the cell type (e.g., 2 h for lymphocytes and bone marrow cells and 4 h for primary fibroblasts). In the case of highly proliferative cells, such as bone marrow, it may be sufficient to add colcemid during hypotonic treatment without the need to culture the cells. Alternatively, to arrest cells in metaphase, nocodazole may be used at a final concentration of 50 μg/mL.
11. In the case of mixed cultures that contain both adherent and nonadherent cells, such as bone marrow, or if it is desirable to include floating cells from the adherent culture in telomere length analysis, aspirate the medium from the original cell culture and keep it to inactivate trypsin used to harvest the adherent cell population.

12. When KCl is used for hypotonic treatment, incubation at 37°C must be reduced to 15–20 min or, alternatively, performed for 25–30 min at room temperature.
13. Saturation levels vary depending on the bit depth at which the images were taken (e.g., 255 for 8-bit and 4095 for 12-bit images). With our system, images are acquired at 8-bit depth.
14. To obtain a linear correlation between fluorescence intensity and exposure time, it is essential to turn off postprocessing options, such as gamma adjustments.
15. Fluosphere slides are prepared by spreading 2 μL of carboxylate-modified Orange Fluospheres (0.2 μm diameter; diluted 1:25–1:50 in fetal calf serum) over the entire surface of a slide and allow them to air-dry. Subsequently, 10 μL of component A of Slow Fade Antifade Kit are added on a coverslip and, avoiding air bubbles, the dry slide is placed upside down on the coverslip. Allow to sit for some minutes until the suspension has spread completely.
16. Intensity values are recorded at several points in the central area and corners of the fluosphere slide. In a homogenous light field, intensity values should not vary by more than 5%. The mean of fluosphere signal intensities, which may suffer variations over time, is used to correct telomere signal intensity values.
17. Alternatively, plasmids that contain cloned telomere sequences of defined length can be used. However, the extrapolation from plasmids with very few and histone-free telomere repeats to telomeric chromatin of up to 150 kb might be problematic. For this reason, it is preferable to calibrate the method with telomeres from cells with defined telomere lengths as determined by TRF.
18. A further alternative to external calibration of the Q-FISH system using cell lines or plasmids with known telomere length is the use of a centromeric probe, which serves as internal control. This method allows calculation of centromere/telomere ratios that are used to compensate for local background fluorescence.
19. Please note that the TFL-TELO software only accepts images in bitmap (BMP) format.
20. Frequently, telomere length values are not following a normal distribution but present a skewed distribution. For skewed distributions, the median or the mode should be used to calculate the "mean" telomere length. Nonetheless, it is very common in the field to use mean values ± standard deviation or standard error for the quantitative description of telomere length

distributions. Note that standard deviations in this type of experiment do not reflect experimental inaccuracy but are an inherent characteristic feature and reflection of the broad and asymmetric distribution of telomere length. In the literature, standard errors, which are very small as a result of the large number of analyzed telomeres, rather than standard deviations, are frequently reported. To calculate whether two telomere distributions are significantly different from each other, the Wilcoxon Rank Sum test can be used.

21. For better results, a confocal microscope can be used to analyze more accurately the Q-FISH interphase nuclei. With the Leica TCS-SP2-AOBS-UV confocal microscope, and by applying optical sample sectioning, all telomeres of an inter- phase nucleus can be captured successfully. In this case, maximal projection intensity can be quantified using the LCS lite Leica software.
22. The DAPI image can be helpful in identifying the morphological structure one wishes to analyze. In some cases, to identify the cell type one wishes to analyze, it may be necessary to use a correlative section stained with histological dyes, i.e., hematoxylin-eosin stain, or to apply the immunostaining Q-FISH technique.
23. For Flow-FISH analysis, isolated cells in suspension, either from cell culture or from tissue disaggregation, can be used.
24. Propidium iodide displays a broad emission spectrum (550–750 nm), which causes considerable interference with the FITC signal from telomeres. This problem is circumvented by maintaining propidium iodide concentrations low (0.1 μg/mL) and by compensation of fluorescence signals between FL2 (propidium iodide) and FL1 (FITC). As alternatives to propidium iodide, a number of commercially available compounds can be used for DNA counter-staining: 0.05–0.1 μg/mL LDS 751, 10 μg/mL 7-AAD, 10 μM Hoechst H 33342, 0.1 μg/mL DAPI, and 0.1-1 μM DRAQ5. For each of these compounds, staining is performed after the last wash step by incubating cells for at least 20 min at room temperature in PBS or FACSFlow, containing 0.1% (w/v) BSA, 10 μg/mL RNAse A, and the DNA counterstain at the final concentration indicated above. In particular, 7-AAD and LDS 751 exhibit less interference with FITC. For more detailed tips related to telomere length determinations by Flow-FISH.

11

Hybridization Techniques

Aging of cells, tissues, or of the whole organism composed of processes that are generally associated with terminal events of their life cycle. Two major theories have been attributed for this syndrome. The first theory, "*programmed senescence*," is based on the belief that aging results from a genetic program, a type of "*biological clock*" that determines the onset and the biochemical events leading to the death of the cell, organ or the whole organism. The second theory holds that aging is a kind of a "*wear and tear*" phenomenon caused mainly by random accumulation of errors due to the generation of unstable molecules known as free radicals.

Most of monocarpic (annuals) plants are known to display "*programmed senescence*," whereas human aging is believed to be predominantly regulated by free-radical damage. Whether the aging and senescence processes represent a wear-and-tear type process of damage accumulation or whether it is, like development, an ordered, programmed process, the molecular and cellular events can be identified and studied by comparing global gene expression in each of the systems.

Leaf senescence is a key developmental step in the life cycle of an annual plant, because the organic material accumulated in the leaf during its growth is remobilized into the developing fruits and seeds during the leaf senescence phase. Although there is extensive physiological and biochemical data on leaf senescence, the molecular events and mechanisms involved in this process are not clearly understood. There is good evidence that leaf senescence, similarly to other controlled developmental processes, is induced and regulated by gene expression of specific set of genes. Thus, it is most likely that a comprehensive study of differential gene expression during senescence will shed light on the molecular events underlying the regulation of this syndrome.

Several techniques aimed at the identification of genes that show enhanced expression during senescence have been published. Among them are: differential screening, subtractive hybridization, differential display, cDNA-AFLP, and suppression subtractive hybridization.

The availability of cDNA microarrays and affymetrix GeneChips has considerably increased the speed by which differential expression genes can be identified in model plant such as *Arabidopsis*. However, lack of microarrays availability for other species or cost limitations makes *suppression subtractive hybridization* (SSH) the method of choice.

Although several subtraction hybridization techniques have been employed for identification of gene products during aging and senescence, the present chapter describes in details one of the latest and most efficient versions of subtractive hybridization.

SSH is a powerful technique enabling researchers to compare two populations of mRNA and identify the differentially expressed genes. Both mRNA populations are used for cDNA preparation: one, designated as the "*tester*," is the cDNA population that contains the specific (preferentially expressed) transcripts and the other is the "*driver*," the reference cDNA population. Tester and driver cDNAs are hybridized, and the resulted hybrid sequences are removed. Consequently, the remaining unhybridized cDNA clones represent genes that are preferentially expressed in the tester. The main advantage of the SSH method, as compared to other subtraction methods, is the selective amplification of the nonabundant, differentially expressed sequences and the suppression of the abundant ones. Thus, rare genes are equally represented as the most abundant genes.

MATERIALS

The following list of components are provided with the PCR-select cDNA subtraction kit (Clontech).

First Strand Synthesis

1. AMV reverse transcriptase (20 U/μL).
2. cDNA synthesis primer (10 μM) 5' $TTTTGTACAAGCTT_{30}\ N_1N$ 3'.
3. 5X first-strand buffer: 250 m*M* Tris-HCl pH 8.5, 40 m*M* $MgCl_2$, 150 m*M* KCl, 5 m*M* dithiothreitol (DTT).

Second-strand Synthesis

1. 20X second-strand enzyme Cocktail: DNA polymerase I, 6 U/μL; RNase H 0.25 U/μL; *Escherichia coli* DNA ligase 1.2 U/μL.

2. 5X second-strand buffer: 500 m*M* KCl, 50 m*M* Ammonium sulfate, 25 m*M* $MgCl_2$, 0.75 m*M* β-NAD, 100 m*M* Tris-HCl pH 7.5, 0.25 mg/mL bovine serum albumin (BSA).
3. T4 DNA polymerase (3 U/μL).

Endonuclease Digestion

1. 10X *Rsa* I restriction buffer: 100 m*M* Bis Tris Propane-HCl pH 7.0, 100 m*M* $MgCl_2$, 1 m*M* DTT.
2. *Rsa* I (10 U/μL).

Adaptor Ligation

1. T4 DNA ligase (400 U/μL; contains 3 m*M* ATP).
2. 5X ligation buffer: 250 m*M* Tris-HCl pH 7.8, 50 m*M* $MgCl_2$, 10 m*M* DTT, 0.25 mg/mL BSA.
3. Adaptor 1 (10 μ*M*): 5' CTAATACGACTCACTATAGGGCTCGA-GCGGCCGCCCGGGCAGGT 3'3' GGCCCGTCCA 5'.
4. Adaptor 2R (10 μ*M*): 5' CTAATACGACTCACTATAGGGCAGC-GTGGTCGCGGCCGAGGT 3'3'GCCGGCTCCA 5'.

Hybridization

1. 4X Hybridization buffer.
2. Dilution buffer: 20 m*M* HEPES-HCl pH 8.3, 50 m*M* NaCl, 0.2 m*M* EDTA pH 8.0.

PCR Amplification

1. PCR Primer 1 (10 μ*M*): 5' CTAATACGACTCACTATAGGGC 3'.
2. Nested PCR primer 1 (10 μM) 5' TCGAGCGGCCGCCCGGGC AGGT 3'.
3. Nested PCR primer 2R (10 μ*M*) 5' AGCGTGGTCGCGGCCGAG GT 3'.
4. PCR control subtracted cDNA.

General Reagents

1. dNTP mix (10 m*M* each dATP, dCTP, dGTP, dTTP).
2. 20X EDTA/Glycogen Mix (0.2 M EDTA; 1 mg/ml glycogen).
3. 4 *M*NH_4OAc.
4. Sterile H_2O.

Additional Materials Required

1. 80% and 96% ethanol.
2. Phenol; chloroform; isoamyl alcohol (25;24;1). Prepare as the following:

(a) Melt phenol.

(b) Equilibrate with as equal volume of sterile TNE buffer (50 m*M* Tris pH 7.5, 150 m*M* NaCl, 1 m*M* EDTA).

(c) Incubate the mixture at room temperature for 2–3 h.

(d) Remove and discard the top layer.

(e) Add an equal volume of chloroform; isoamyle alcohol (24;1) to the remaining layer. Mix thoroughly. Remove and discard the top layer.

(f) Store the bottom layer of Phenol; chloroform; isoamyl alcohol at 4°C away from light for a maximum of 2 wk.

3. Chloroform; isoamyle alcohol (24;1).

4. 10X PCR buffer.

5. dNTP Mix for PCR.

6. 50X TAE electrophoresis buffer: 242 g Tris base, 57.1 mL Glacial acetic acid, 37.2 g $Na_2EDTA \cdot 2H_2O$.

METHODS

First, cDNA is synthesized from 0.5–2 μg of poly A^+ RNA from the two types of tissues or cells being compared, i.e., the tester, which is the cDNA population that contains the specific (preferentially expressed) transcripts and the driver which is the reference cDNA population. The tester and driver cDNAs are then digested with *Rsa* I, a restriction enzyme that yields blunt ends. The tester cDNA is then subdivided into two portions, and each is ligated with a different cDNA adaptor. The ends of the adaptor do not have a phosphate group, so only one strand of each adaptor attaches to the 5' ends of the cDNA. The two adaptors have stretches of identical sequence to allow annealing of the PCR primer once the recessed ends have been filled in.

Two hybridizations are then performed. In the first, an excess of driver is added to each sample of tester. The samples are then heat denatured and allowed to anneal, generating the type a,b,c, and d molecules in each sample. The concentration of high- and low-abundant sequences is equalized among the type a molecules because reannealing is faster for the more abundant molecules as a result of the second-order kinetics of hybridization. At the same time, the single-strand type a molecules are significantly enriched for differentially expressed sequences, as cDNAs that are not differentially expressed form type c molecules with the driver.

During the second hybridization, the two primary hybridization samples are mixed together without denaturing. Now, only the remaining

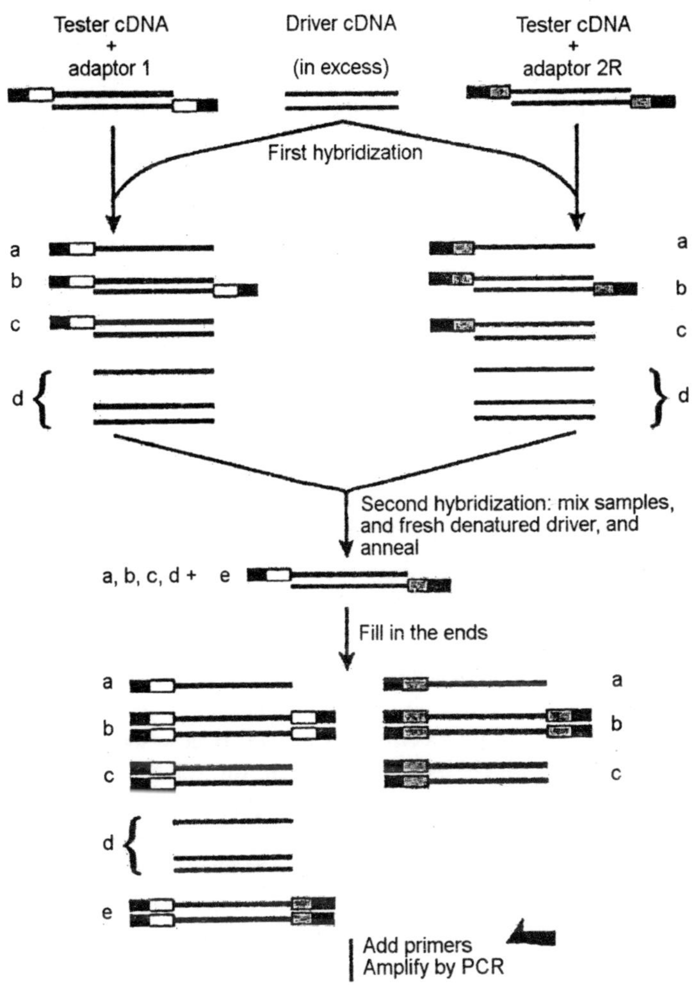

Fig. 11.1. Principles of suppression substractive hybridization.

equalized and subtracted ss tester cDNAs can reassociate and form new type of hybrids. These new hybrids are ds tester molecules with different ends, which corresponded to the sequences of Adaptor 1 and 2R. Fresh denatured driver cDNA is added (again without denaturing the subtraction mix) to further enrich fraction *e* for differentially expressed sequences. After filling in the ends by DNA polymerase, the type *e* molecules—the differentially expressed tester sequences—have different annealing sites for the nested primers on their 5' and 3' ends.

The entire population of molecules is then subjected to PCR to amplify the desired differentially expressed sequences. During PCR, type *a* and *d* molecules are missing primer annealing sites, and thus cannot be amplified. Because of the suppression PCR effect, most type *b* molecules form a pan-like structure that prevents their exponential amplification. Type *c* molecules have only one primer annealing site and can only be amplified linearly. Only type e molecules, which have two different adaptors, can be amplified exponentially. These are the equalized, differentially expressed sequences.

Next, a second PCR amplification is performed using nested primers to further reduce any background PCR products and to enrich differentially expressed sequences. The cDNAs can then be directly inserted into a T/A cloning vector. Alternatively, the *Not* I site on Adaptor I and the *Eag* I site on Adaptor 2R can be used for blunt-end cloning. It is important to confirm that individual clones indeed represent differentially expressed genes. This is typically accomplished by probing Northern blots with randomly selected, subtractive clone, or by differential screening using dot blot arrays.

General Considerations

We have good experience using the method of PCR-select cDNA subtraction kit of Clonetech, which is based on the method of Diatchenko. Thus, the protocol of this method is described. The total RNA extraction, which is a critical step for mRNA and cDNA preparation prior to the subtraction, has been prepared for plant tissues. Thus it is recommended that one use the appropriate RNA isolation method for cells and tissues of the specific studied species.

Wear gloves to protect RNA and cDNA samples from degradation by nucleases. Use aerosol-free pipet tips to measure small volumes and sterile disposable pipets for larger volumes.

First-Strand cDNA Synthesis

1. For each tester and driver, combine the following components in a sterile 0.5-mL microcentrifuge tube. (Do not use a polystyrene tube).

   Poly A RNA (2 μg) 2–4 μL

   cDNA synthesis primer (10 *μM*) 1 μL

   If needed, add sterile H_2O to a final volume of 5 μL. Mix contents and spin the tubes briefly in a microcentrifuge.
2. Incubate the tubes at 70°C in a thermal cycler for 2 min.

3. Cool the tubes on ice for 2 min.
4. Briefly centrifuge the tubes.
5. Add the following to each reaction tube:

5X First-strand buffer	2 μL
dNTP Mix (10 μM each)	1 μL
Sterile H_2O	1 μL
AMV reverse transcriptase (20 U/μL)	1 μL

6. Gently vortex and briefly centrifuge the tubes.
7. Incubate the tubes at 42°C for 1.5 h in an air incubator.
8. Place the tubes on ice to terminate first-strand cDNA synthesis.

Second-strand cDNA Synthesis

Perform the following procedure with each first-strand taster and driver.

1. Add the following components to the first-strand synthesis reaction tubes (containing 10 μL):

Sterile H_2O	48.4 μL
5X second-strand buffer	16.0 μL
dNTP Mix (10 *μM*)	1.6 μL
20X second-strand enzyme cocktail	4.0 μL

2. Mix contents and briefly spin the tubes. The final volume should be 80 μL.
3. Incubate tubes at 16°C for 2 h.
4. Add 2 μL (6 U) of T4 DNA Polymerase. Mix contents well.
5. Incubate the tubes at 16°C for 30 min.
6. Add 4 μL of 20X EDTA/Glycogen Mix to terminate second-strand synthesis.
7. Add 100 μL of phenol:chloroform:isoamyl alcohol (25:24:1).
8. Vortex thoroughly, and centrifuge the tubes at 14,000 rpm for 10 min at room temperature.
9. Carefully remove the top aqueous layer and placed in a clean (sterile) 0.5-mL microcentrifuge tube. Discard the interphase and lower phase.
10. Add 100 μL of chloroform:isoamyl alcohol (24:1) to the aqueous layer.
11. Repeat steps 8 and 9.
12. Add 40 μL of 4 *M* NH_4OAc and 300 μL of 95% ethanol. Proceed immediately and do not store tubes at –20°C.

13. Vortex thoroughly and centrifuge the tubes at 14,000 rpm for 10 min at room temperature.
14. Remove the supernatant carefully.
15. Overlay the pellet with 500 μL of 80% ethanol.
16. Centrifuge the tubes at 14,000 rpm for 10 min.
17. Remove the supernatant.
18. Air-dry the pellet for about 10 min to evaporate residual ethanol.
19. Dissolve precipitate in 50 μL of H_2O.
20. Transfer 6 μL to a fresh microcentrifuge tube. Store this sample at -20°C until after *Rsa* I digestion for agarose gel electrophoresis to estimate yield and size range of ds cDNA products synthesized.

Rsa *I Digestion*

Perform the following procedure with each experimental ds tester and driver cDNA. This step generates shorter, blunt ended ds cDNA fragments which are optimal for subtraction and necessary for adaptor ligation.

1. Add the following reagents into the tube:

ds cDNA	43.5 μL
10X *Rsa* I restriction buffer	5.0 μL
Rsa I (10 U/μL)	1.5 μL

2. Mix by vortexing and centrifuging briefly.
3. Incubate at 37°C for 1.5 h.
4. Set aside 5 μL of the digest mixture to analyze the efficiency of *Rsa* I digestion.
5. Add 2.5 μL of 20X EDTA/glycogen mix to terminate the reaction.
6. Add 50 μL of phenol:chloroform:isoamyl alcohol (25:24:1).
7. Vortex thoroughly.
8. Centrifuge the tubes at 20,000*g* for 10 min to separate phases.
9. Remove the top aqueous layer and place in a clean 0.5-mL tube.
10. Add 50 μL of chloroform:isoamyl alcohol (24:1) and vortex thoroughly.
11. Centrifuge the tubes at 14,000 rpm for 10 min to separate phases.
12. Remove the top aqueous layer and place in a clean 0.5-mL tube.
13. Add 25 μL of 4 *M* NH_4OAc and 187.5 μL of 95% ethanol.
14. Vortex the mixture thoroughly.
15. Centrifuge the tubes for 20 min at 20,000*g* at room temperature.
16. Remove the supernatant carefully.

17. Gently overlay the pellet with 200 μL of 80% ethanol.
18. Centrifuge at 20,000*g* for 5 min.
19. Carefully remove the supernatant.
20. Air-dry the pellets for 5–10 min.
21. Dissolve the pellet in 5.5 μL of H_2O and store at -20°C. These 5.5-μL samples of ***Rsa*** I-digested cDNA will serve as your experimentally driver cDNA. In the next section, these samples will be ligated with adaptors to create your tester cDNA.
22. Chcck your *Rsa* I-digested cDNA.

Adaptor Ligation

The forward subtraction experiment designed to enrich for differentially expressed sequences present in polyA RNA sample 1 (cDNA 1, tester) but not polyA RNA sample 2 (cDNA 2, driver). The reciprocal (reverse) subtraction, in which cDNA 2 serves as tester and cDNA 1 serves as driver. The result is two subtracted cDNA population: the forward-subtracted cDNA contains sequences that are

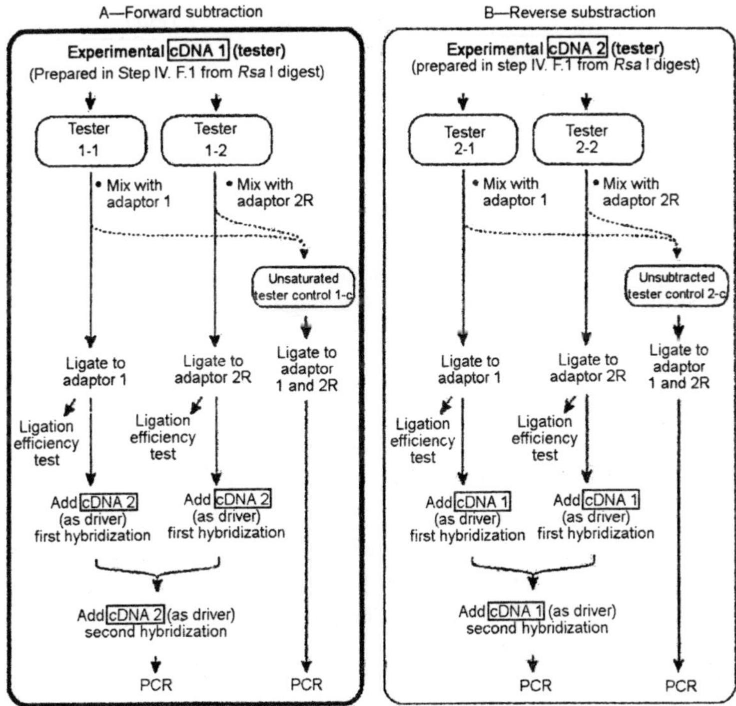

Fig. 11.2. Preparing adaptor-ligated tester cDNAs for hybridization and PCR.

specific to sample 1, and the reverse-subtracted cDNA contains sequences that are specific to sample 2.

To perform subtractions in both directions, you will need to prepare tester cDNA corresponding to each of your polyA+ RNA samples.

Each tester cDNA is aliquated into two separate tubes: one aliquot is ligated with Adaptor 1 (Tester 1-1, 2-1), and the second is ligated with Adaptor 2R (Tester 1-2, 2-2). After the ligation reactions are set up, portions of each tester tubes are combines so that the cDNA is ligated with both adaptors (Unsubtracted tester control 1-c, 2-c). The Unsubtracted tester control cDNA serves as a positive control for ligation, and later serves as a negative control for subtraction.

Note: adaptors will not be ligated to the driver cDNA.

1. Dilute 1 μL of each *Rsa* I-digested experimental cDNA, with 5 μL of distilled water. Prepare your adaptor-ligated tester cDNA.

Table 11.1. Setting up the ligation reaction

Component	*Tube 1* *Tester 1-1 (μL)*	*Tube 2* *Tester 1-2 (μL)*
Dilute tester cDNA	2	2
Adaptor 1 (10 *μM*)	2	—
Adaptor 2R	—	2
Master Mix	6	6
Final volume	10	10

2. Prepare a ligation Master Mix by combining the following reagents in a 0.5-mL microcentrifuge tube. To ensure that you have sufficient Master Mix, prepare enough for all ligation plus one additional reaction.

	Per reaction
Sterile H_2O	3 μL
5X ligation buffer	2 μL
T4 DNA Ligase (400 U/μL)	1 μL

The ATP required for ligation is in the T4 DNA ligase.

3. For each experimental tester cDNA, combine the reagents in Table 11.1 in the order shown in 0.5-mL microcentrifuge tubes. Pipet mixture up and down to mix thoroughly. Use the same setup for Tester 2-1 and 2-2.
4. In a fresh microcentrifuge tube, mix 2 μL of Tester 1-1 and 2 μL of Tester 1-2. After ligation is completed, this will be the Unsubtracted tester control 1-c. Do the same for each additional

experimental tester cDNA. After ligation, approx one- third of the cDNA molecules in the Unsubtracted tester control tube will bear two different adaptors.

5. Centrifuge tubes briefly, and incubate at 16°C overnight.
6. Stop ligation reaction by adding 1 μL EDTA/glycogen mix.
7. Heat samples at 72°C for 5 min to inactivate the ligase.
8. Briefly centrifuge the tubes. Preparation of the experimental adaptor-ligated Tester cDNAs and your Unsubtracted tester control is now complete.
9. Remove 1 μL from each Unsubtracted tester control (1-c) and dilute into 1 mL of H_2O. These samples will be used for PCR.
10. Store samples at –20°C.

First Hybridization

In the following procedure, an excess of driver cDNA is added to each tester cDNA, and the samples are heat denatured and allowed to anneal. The remaining ss cDNAs (available for the second hybridization) are dramatically enriched for differentially expressed sequences, as nontarget cDNAs present in the tester and driver cDNA form hybrids.

Important: make sure that prior to the hybridization, the 4X Hybridization buffer has been allowed to warm to room temperature for at least 15–20 min.

Table 11.2. Setting up the first hybridization

Component	*Hybridization sample 1 (µL)*	*Hybridization sample 2 (µL)*
Rsa I-digested driver cDNA	1.5	1.5
Adaptor 1-ligated Tester 1-1	1.5	–
Adaptor 2R-ligated Tester 1	–	1.5
4X hybridization buffer	1.0	1.0
Final volume	4	4

Be sure that there is no visible pellet or precipitate before using the buffer. If necessary, heat the buffer at 37°C for 10 min to dissolve any precipitate.

1. For each of the experimental subtraction, combine the reagent in Table 11.2 in 0.5-mL tubes in the order shown.
2. Overlay samples with one drop of mineral oil and centrifuge briefly.
3. Incubate samples in a thermal cycler at 98°C for 1.5 min.
4. Incubate samples at 68°C for 8 h, and then proceed immediately. Use the same setup for Tester 2-1 and 2-2.

Second Hybridization

The two samples from the first hybridization are mixed together, and fresh denatured driver DNA is added to further enrich for differentially expressed sequences. New hybrid molecules are formed which consist of differentially expressed cDNAs with different adaptors on each end.

Important: Do not denature the primary hybridization samples at this stage. Also, do not remove the hybridization samples from the thermal cycler for longer than is necessary to add fresh driver. Repeat the following steps for each experimental tester cDNA.

1. Add the following reagents into a sterile tube:

Driver cDNA	1 μL
4X hybridization buffer	1 μL
Sterile water	2 μL

2. Place 1 μL of this mixture in a 0.5 mL microcentrifuge tube and overlay it with one drop of mineral oil.
3. Incubate in a thermal cycler at 98°C for 1.5 min.
4. Remove the tube of freshly denatured driver from the thermal cycler. Use the following procedure to simultaneously mix the driver with hybridization samples 1 and 2.

   This ensures that the two hybridization samples mix together only in the presence of freshly denatured driver.

   (a) Set a micropipettor at 15 μL.
   (b) Gently touch the pipet tip to the mineral oil/sample interface of the tube containing hybridization sample 2.
   (c) Carefully draw the entire sample partway into the pipet tip. Do not worry if a small amount of mineral oil is transferred with the sample.
   (d) Remove the pipet tip from the tube, and draw a small amount of air into the tip, creating a slight air space below the droplet of sample.
   (e) Repeat steps b–d with the tube containing the freshly denatured driver. The pipet tip should now contain both samples separated by a small pocket of air.
   (f) Transfer the entire mixture to the tube containing hybridization sample 1.
   (g) Mix by pipetting up and down.
5. Briefly centrifuge the tube if necessary.

6. Incubate reaction at 68°C overnight.
7. Add 200 μL of dilution buffer to the tube and mix by pipetting.
8. Heat in a thermal cycler at 68°C for 7 min.
9. Store at -20°C.

PCR Amplification

Differentially expressed cDNAs are selectively amplified during the reaction as described in this section. Prior to thermal cycling, you will fill in the missing strands of the adaptors by a brief incubation at 75°C. This creates the binding site for PCR primer 1. In the first amplification, only ds cDNAs with different adaptor sequences on each end are exponentially amplified. In the second, nested PCR is used to further reduce background and to enrich for differentially expressed sequences. The PCR should be performed for: (1) forward-subtracted experimental cDNA, (2) The Unsubtracted tester control (1-c), (3) the reversed-subtracted experimental cDNA, (4) the reversed Unsubtracted tester control (2-c), (5) the PCR control subtracted cDNA (provided with the kit). The PCR control subtracted cDNA provides a positive PCR control and contains a successfully subtracted mixture of *Hea* III-digested ΦX174 DNA. It is also recommended that one perform a positive and negative controls for the enzyme used for the reaction.

A hot start must be used, as follows: (1) prepare the primary PCR Master Mix without *Taq* polymerase; (2) mix PCR samples and heat the reaction mix to 75°C for 1 min; (3) quickly add the necessary amount of *Taq* polymerase; (4) incubate the reaction at 75°C for 5 min; (5) perform PCR as described at step 8 below.

Table 11.3. Preparation of the primary PCR master mix

Reagent	*Amount per reaction (μL)*
Sterile H_2O	19.5
10X PCR reaction buffer	2.5
dNTP mix (10 m*M*)	0.5
PCR primer 1 (10 μ*M*)	1.0
Total volume	23.5

1. Prepare the PCR templates:
   (a) Aliquot 1 μL of each diluted cDNA (each subtracted sample and the corresponding diluted Unsubtracted tester control) into an appropriately labeled tube.
   (b) Aliquot 1 μL of the PCR control subtracted cDNA (provided in the kit) into an appropriately labeled tube.

2. Prepare a Master Mix for all the primary PCR tubes plus one additional tube. For each reaction planned, combine the reagents in **Table 3** in the order shown.
3. Mix well by vortexing, and briefly centrifuge the tube.
4. Aliquot 23.5 μL of Master Mix into each of the reaction tubes prepared in step 1.
5. Overlay with 50 μL of mineral oil.
6. Incubate the reaction mix in a thermal cycler at 75°C for 1 min. Add 0.5 μL of *Taq* Polymerase to each tube.
7. Incubate the reaction mix in a thermal cycler at 75°C for 5 min to extend the adaptors. (Do note remove the samples from the thermal cycler.)
8. Immediately commence thermal cycling:

   31 cycles:

94°C	30 s
66°C	30 s
72°C	1.5 min

9. Analyze 8 μL from each tube on a 2 % agarose/EtBr gel run in 1X TAE buffer.
10. Dilute 3 μL of each primary PCR mixture in 27 μL of H_2O.
11. Aliquot 1 μL of each diluted primary PCR product mixture from step 10 into an appropriately labeled tube.
12. Prepare Master Mix for the secondary PCRs plus one additional reaction by combining the reagents in Table 4 in the order shown.
13. Mix well by vortexing, and briefly centrifuge the tube.
14. Aliquot 23.5 μL of Master Mix into each of the reaction tubes prepared in step 11.
15. Overlay with 50 μL of mineral oil.
16. Immediately commence thermal cycling:

   94°C 30 s

   68°C Hold temperature and add 0.5 μL of *Taq* polymerase to each tube

   68°C 30 s

   16 cycles:

94°C	30 s
68°C	30 s
72°C	1.5 min

Table 11.4. Preparation of the secondary PCR master mix

Reagent	*Amount per reaction (μL)*
Sterile H_2O	18.5
10X PCR reaction buffer	2.5
dNTP mix (10 m*M*)	0.5
Nested PCR primer 1 (10 μ*M*)	1.0
Nested PCR primer 2R (10 μ*M*)	1.0
Total volume	23.5

17. Analyze 8 μL from each tube on a 2% agarose/EtBr gel run in 1X TAE buffer.
18. Store reaction products at –20°C.

The PCR mixture is now enriched for differentially expressed cDNAs. In addition, differentially expressed transcripts that varied in abundance in the original mRNA sample should now be present in roughly equal proportions.

Cloning and Northern Blot Analysis

The cDNA products generated by the SSH are cloned into a T/A cloning vector and the plasmids are transformed into *E. coli*. PCR analyses for each of the colonies are performed, and the PCR products are usually confirmed by separation on a 1% agarose gel. The relevant colonies are being grown for plasmid isolation and also colonies samples kept frozen at –80°C for further research. DNA sequence analysis is an essential step for gene identification and is being done following plasmid isolation of the individual colonies.

Research aimed at the identification of genes associated with leaf senescence of *Arabidopsis* has been carried out using the SSH approach and revealed hundreds of senescence-related genes. Northern blot analyses have indicated the existence of various gene groups that display different temporal expression patterns such as those characteristic to early expressed or late expressed genes.

Notes

1. To monitor the progress of cDNA synthesis, dilute 1 μL of [α^{32}P]dCTP (10 mCi/mL) with 9 μL of H_2O, and replace H_2O above with 1 μL of the diluted label.
2. If you used [α^{32}P]dCTP, check for the pellet using a Geiger counter.
3. Proceed immediately with precipitation. Do not store tubes at –20°C. Prolonged exposure to this temperature can precipitate unwanted salts.

4. Electrophorese 2.5 μL of undigested, ds cDNA and 5 μL of *Rsa* I-digested cDNA on a 1% agarose/EtBr gel in 1X TAE buffer. Compare the results side-by-side. cDNA derived from polyA$^+$ RNA appears as a smear from 0.5–20 kb. Bright bands correspond to abundant mRNAs or rRNAs. After *Rsa* I digestion, the average cDNA size is smaller (0.1–2 kb). If the size distribution of your sample is not reduced after *Rsa* I digestion, repeat phenol/chloroform extraction, ethanol precipitation, and digestion. If [α^{32}P]dCTP was used, it is also possible to expose the gel to a phosphoimager intensifying screen overnight to visualize the results.
5. Through the rest of the procedure, be sure to label the tubes using the nomenclature described in the manual. Labeling the tubes of intermediate products with the step number in which they were created may prove helpful as well.
6. Samples may hybridize for as little as 6 h, or as much as 12 h. Do not let the incubation exceed 12 h.
7. The experimental primary PCR subtraction products usually appear as a smear from 0.2–2 kb, with or without some distinct bands. If you cannot visualize a smear, repeat the primary PCR with some modification like addition of three more cycles, or reduce the annealing temperature to 64°C. Sometimes it takes few repeats until it is successful.
8. The experimental subtracted samples usually look like smears with or without a number of distinct bands. However, if you can see only, or very clear bands, it can imply that few clones had taken over the library. You must repeat the primary and secondary PCR; otherwise most of the colonies will contain redundant sequences.

12

MAPPING IN AGING SYSTEM

Animal models and centenarian studies have shown that aging is the result of complex interactions of genes and environment. The best universal biomarker and also a quantitatively measurable biomarker for aging and immunosenescence in higher species is thymic involution. The shrinkage of thymus was reported more than 70 yr ago by Boyd. Thymic involution can occur in normal, unstressed, unmanipulated mice as a consequence of aging. Thymic involution also meets all basic four criteria proposed by Strehler to describe the fundamental age-related changes, including reduction in reduct function, gradual progression, occurs as an intrinsic effect, and most importantly, is a universal phenomenon shared by many different species. The impact of thymic involution is even more significant because we have observed that BXD recombinant inbred (RI) mouse strains that exhibited a slower thymic involution rate with age also exhibited a better resistance and *cytotoxic T lymphocyte* (CTL) response in a tumor model. However, the genetic factors that influence age-related thymic involution are not well understood. We have shown that the process of thymic involution is highly amenable to mathematical modeling and multi-trait mapping approaches. In this chapter, we will use thymic involution as an example of an approach to determine the QTL mapping to understand aging systems. It should be pointed out that the purpose of this chapter is to introduce the available tools and some of the basic software usage methods that can be utilized for the QTL analysis. However, to fully appreciate the essence of complex genetic linkage analysis in rodent populations, it will be necessary for the readers to refer to other references for detailed background information.

For genetic linkage studies, the underlying assumption is that the genetic factors that influence immune senescence can be dissected by understanding and accurately measuring key quantitative traits that have heritable variation among individuals or entire strains. A quantitative trait is ideally one that is easy to measure in a highly reproducible way and best reflects the aging phenotype under study. One approach to identifying such quantitative traits is to first define the mechanistic framework by defining the hierarchy of the critical initiating events and the major modifying events. In order to identify the causes of thymic involution, it is necessary to define involution accurately and informatively. This is particularly true in the case of selecting the parameters to be used in QTL analysis, which also requires that methods of analysis of the thymus are available that permit screening of a large number of mice.

Classically, involution can be defined in terms of changes in the whole thymus: its weight, its size (volume), and/or the total number of thymocytes. This information is relatively easy to obtain and is informative in that it conforms to the classical definition of the process. However, thymic involution does not occur at a constant rate, and can be considered either as a negative growth trajectory or a finite set of measurements. Use of genetically identical RI strains allows repeated measurements at different ages in genetically identical individuals. This can be repeated for each RI strain within a set to yield quantitative traits that can be used for a variety of studies, including complex trait analysis and QTL mapping.

The use of RI strains of mice is especially well suited to study this or similar age-related processes that require sacrifice of genetically identical mice at different times to assess the trait under investigation. Such studies require analysis of a proper set of different strains of mice, each strain being genetically identical, at different times. A similar analysis could be carried out comparing other RI strains of mice that have been genetically mapped. A second approach, not discussed here, would be the use of a whole panel of F2 mice generated by intercrossing the (B6 × D2)F 1 mice. A disadvantage of this approach is that each mouse would be genetically different and would have to be individually genotyped, and it would not be possible to follow thymic involution in a defined genotype at different times.

Based on our experience with genetic linkage analysis of the thymic involution, this can be accomplished by a multifaceted approach and is best accomplished through the following strategies:

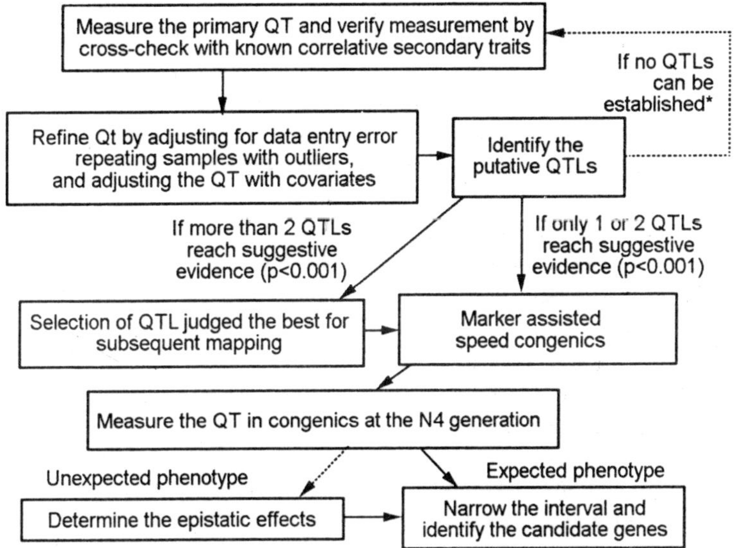

Fig. 12.1. Summary of strategies that can be used to determine the quantitative trait locus mapping in aging systems.

1. Identification of the phenotypes of interest, including validation and power analysis, and identification of age-related changes in phenotypes associated with thymic involution.
2. Analyses of the QTL associated with the defined phenotypes.
3. Confirmation of the genetic effects of the loci on the thymic involution phenotype.

This chapter describes an approach to understanding the genetic basis of thymic involution. However, the same approach can be applied to any quantitative trait that can be measured at different ages within RI stains of mice, and that exhibits a hereditary difference within the chosen RI sets.

Materials

Mice

The development of the panel of C57BL/6 × DBA/2 (BXD) RI mouse strains provides the foundation that enables this approach in that it provides: (a) a well characterized system that permits resampling of thymii from genetically identical animals with significantly different rates of involution, life spans, and the turnover rate of the bone marrow progenitor cells; and (b) a genetic model that will permit determination of the relative significance of age-associated changes in the thymus.

The BXD recombinant inbred mouse strains are constructed by crossing two commonly used inbred strains (B6 and D2) to produce a (B6 × D2)F1 strain, followed by 20 or more consecutive generations of brother × sister mating from this point on to obtain the different RI strains of mice. The generated RI set is homozygous at each locus (either b/b or d/d for the BXD RI set) throughout the entire genomes. Because unlinked loci assort randomly in the F2 generation, parental and recombinant allelic combinations occur with approximately equal frequency among a set of RI strains. The BXD strains 1 through 32 were produced by Benjamin A. Taylor starting in the late 1970s. BXD33 through BXD42 were also produced by Taylor, but from a second set of crosses initiated in the early 1990s. This set of RI strains is a remarkable resource because many of these strains have been extensively phenotyped for hundreds of interesting traits over a 25-yr period. A significant advantage of this RI set is that the two parental strains (B6 and D2) have both been extensively sequenced and are known to differ at approx 1.8 million single-nucleotide polymorphisms (SNPs). Thus, in combination with the BXD RI strains genotyping data and the parental B6 and D2 genetic polymorphism data, it is possible to determine the derivation of the polymorphic genes throughout the genomes of the BXD RI strains of mice. Coding variants (mostly SNPs and insertion-deletions) that may produce interesting phenotypes can be rapidly identified in this particular RI set. Each locus has a particular pattern of inheritance called the strain distribution pattern (SDP). Comparisons are made between SDPs. A significant excess of parental genotypes, with respect to two SDPs, signals the possibility of genetic linkage.

The BXD 1 to 42 strains of mice are available from the Jackson Laboratory. To enhance the power of genetic linkage analysis, BXD43 through BXD100 were produced by Williams and colleagues in the late 1990s and early 2000s using advanced intercross progeny. These strains are also available, either from LL and RW or from the Oak Ridge National Laboratory mouse user facility.

Determination of Quantitative Trait

A quantitative trait is a characteristic or condition that is influenced by gene(s), and in a more complicated situation, by the interactions of genes and environment. To minimize the measurement error due to observation-induced changes in the trait, analogous to the classical Heisenberg Uncertainty Principle in quantum mechanics, an ideal quantitative trait would be a variable that can be determined noninvasively, such as the color or weight of the mouse. Noninvasive analysis or minimally invasive analysis such as immune phenotyping or serological analysis at different ages is also a use for a quantitative trait. However, the present example will illustrate the use of a quantitative trait that requires sacrifice of the mouse. For demonstration purposes, we used the age-related decline of thymocyte count as the primary quantitative trait for the analysis. It is therefore desirable to have large numbers of genetically identical mice, such as the use of RI strains of mice.

The second aspect of a quantitative trait is that it should be heritable, and should exhibit differences within the genotype of the set of mice being studied. One indicator that a quantitative trait will exhibit genetic differences is if the trait is different in the parental strains. Interestingly, this is not essential, because segregation of genes that can enhance or inhibit the trait being analyzed in different offspring may be more pronounced within the members of the RI set that observed in the parents. Therefore, even when no significant difference in the quantitative trait is observed between the parental strains, differences of the quantitative trait can still be observed among the RI strains derived from them. This is because two strains with the same phenotype can still vary genetically at QTL due to complementary patterns of positive and negative allelic effects at QTL elsewhere in the genome.

BXD RI Mice Genotyping Database

Genotyping datasets for all BXD strains (BXD1 through BXD 100) are readily available for genetic linkage analysis. This information can be retrieved from the following websites:

1. GeneNetwork. Carefully error-checked consensus genotypes are built into GeneNetwork of the BXDs and most other genetic reference populations of mice.
2. SNP Map Table: C57BL/6J v.s. DBA/2J from the Mouse SNP Data Database maintained by the Whitehead Institute/MIT Center for Genome Research.

Genetic Linkage Analysis

Genetic linkage analysis can be used to identify regions of the genome that contain genes that predispose to the observed quantitative trait, leading to identification of QTLs. A significant QTL means that different genotypes at a polymorphic marker locus are associated with different trait values. Linkage is determined by the log of odds (LOD) scores or likelihood ratio statistics (LRS). To calculate a LOD score or an LRS score for a selected quantitative trait, several programs with insightful tutorials have been developed.

WebQTL

WebQTL is a module built into GeneNetwork for mapping quantitative traits. It incorporates several large transcriptome and phenotype databases and has assembled and curated about 25 yr of published legacy data that includes well over 1000 classical system-level phenotypes in six different sets of RI strains of mice and rats (BXD, AXB/BXA, CXB, BXH, LXS, and HXB/BXH recombinant inbred strains). The largest data sets are transcriptome surveys (Affymetrix and Agilent platforms) of several different tissues and cell types across 30 to 80 strains of mice. Each RI and phenotype data set is accompanied by carefully error-checked genetic maps that can be used to locate modifier genes and quantitative trait loci (QTLs) that often produce variation in phenotypes and disease susceptibility.

Map Manager QTX

Map Manager QTX has a well designed, almost intuitive, interface. Map Manager QTX is available for Macintosh and Windows OS. It is one of the few programs that computes genetic maps requiring only that the user enters the strain of the mouse being analyzed. To take full advantage of the permutation test that is built into Map Manager QTX, this program should be installed on a computer with a fast processor. Map Manager QTX is accompanied by an informative tutorial and manual that can be downloaded and used off line. It is also recommended that the user reads a review by Tanksley on QTL mapping strategies and their limitations.

Other useful QTL mapping softwares

QTL Cartographer is a highly capable mapping program—one that may be particularly suitable for those with a background in UNIX and who are comfortable with advanced statistical analysis.

Windows QTL Cartographer is a command-line sibling and a relatively more user friendly version of QTL Cartographer. This program includes a powerful graphic tool for presenting mapping results and can import and export data in a variety of formats and provide a graphical interface to QTL Cartographer's features. The following statistical methods can be analyzed using Window QTL cartographer: single-marker analysis, interval mapping, composite interval mapping, Bayesian interval mapping, multiple interval mapping, multiple trait analysis, and Categorical trait analysis.

The QTL Cafe links to a Java applet that will help the user carry out an online QTL analysis. The user will need Netscape Navigator 4.05 or higher and will need to have three data files before starting: a map file with information on the names and locations of markers, a file with genotypes, and a file with the case IDs and trait values.

MAPMAKER/QTL is a useful program that implements a maximum likelihood method to map QTLs.

R/qtl is a program developed by Karl W. Broman in Johns Hopkins University. The current version of R/qtl includes facilities for estimating genetic maps, identifying genotyping errors, and performing single-QTL genome scans and two-QTL, two-dimensional genome scans, by interval mapping (with the EM algorithm), Haley-Knott regression, and multiple imputation. All of this may be done in the presence of covariates (such as sex, age or treatment). Furthermore, the code is written so that new methods can be readily implemented. Computationally intensive algorithms were coded in C, whereas the data manipulation and graphics functions were coded in the R language. R/qtl accepts input in a variety of formats and is available for Windows, Unix and Mac OS.

METHODS

Harvest of Thymus

The thymus in the mouse consists of two lobes, and lies on the median line of the vertebral column, close to the base of the heart. A smooth thin capsule covers the two lobes. The lobes have a parenchymatous consistency and white color. In normal conditions, the thymus of an adult mouse of three months of age has an average weight of 30 to 40 mg.

1. Set up a sterile plastic petri dish containing about 5–10 mL of cold, sterile phosphate- buffered saline (PBS) or RPMI1640 medium before starting to remove the thymus from the mouse.
2. Anesthetize and sacrifice the mouse according to the institutional recommended protocol.
3. Place the mouse on its back on a clean surface. Wipe the abdomen and upper abdomen side thoroughly with 70% alcohol to prevent fur contamination.
4. Using one set of sterile forceps and scissors, carefully trim the fur and skin.
5. Using the second set of sterile forceps and scissors, make a midline abdominal incision.

6. Using the same (second) set of sterile scissors to cut the ribs and expose the thoracic cavity. Locate the thymus anterior to the heart and attached to the posterior thoracic well.
7. Remove the thymus by raising the inferior aspect of the two lobes and dissect the thymus from the thoracic wall. Trim away and dispose of excess fatty or connective tissue.
8. To include secondary phenotypes, each thymus is weighed and photographed using a digital camera, and place the two lobes of thymus into the Petri dish.

Determination of Age-Related Thymic Involution Curve in Parental Strains of Mice

1. It is possible to carry out QTL analysis on multiple individual measurements of thymus size at different ages, or to generate a thymic involution curve that captures the age-related change. A thymic involution curve can be generated by plotting the total thymocyte count from the parental strains of mice (B6 and D2) against the age (days) of mice at the time of sacrifice.
2. Regression and statistical analyses is performed using the SPSS statistical package. Different equations, including a linear, logarithmic, inverse, quadratic, cubic, power, S curve, and exponential curve, were used to fit the data and select the functional form that, across all strains, yields the best value of Akaike's information criterion (AIC), a well known measure of model fit relative to model parsimony. Our result shows that the rate of thymic involution can be best fit for both strains of mice with a negative exponential curve, which shows the highest correlation R^2 values compared to other regression curves and can be defined by $N(t) = \beta_0 * \exp(-\beta_1 t)$, where N(t) is the total count or number of thymocytes as a function of age in days (t)(3). The negative exponential curve has previously described as the best curve to fit thymic involution in humans and ICR mice and age-related changes in other biological variables. We have also applied this regression model to data published by Sempowski et al. and have found that this is also the best curve fit model for total thymocyte count ($R^2 = 0.943$, $p = 0.001$) over age in BALB/c mice.

Determination of Secondary Traits that Influence Age-Related Thymic Involution

1. The following secondary phenotypes, thymus size, total thymus weight, and accumulation of $CD44^+CD25^-$ $CD4^-CD8^-$ double-negative (DN)1 thymocytes were used to compare with the data

obtained for the primary trait. This can be compared by plotting the thymocyte count from each strain at each age group against the secondary phenotypes. Counting should be repeated for samples that show low correlation between measurements of the primary traits of interest and those of the secondary reference traits.

2. A more stringent thymocyte phenotype analysis must be performed to ensure that the single-cell suspension obtained for counting contains mostly thymocytes but not peripheral T-cells. For example, we have noticed that single cell suspensions obtained from the thymus of aged autoimmune BXD2 mice contain a large percentage cells that are $CD4^+$ or $CD8^+$ single-positive T-cells and $B220^+$ B-cells. For this reason, the data points obtained from the BXD2 strain of mice were excluded for the QTL analysis.
3. The procedures determine the thymic involution slope can be then applied to determine the $-\beta_1$ value in each strain of BXD RI strain of mice.

Genetic Network Analysis Using WebQTL

The following section describes the genetic linkage analysis using the WebQTL method with the $-\beta_1$ value as the quantitative trait in BXD RI strain.

1. Enter the WebQTL data submission website: http://www.genenetwork.org/cgibin/WebQTL.py.
2. Select the cross or recombinant inbred set from the menu by selecting "BXD" in the box.
3. Enter the quantitative trait data into the WebQTL data sheet either from a file or by pasting or typing multiple values into the provided area. As an example for using both methods, the data for the BXD thymic involution slope in the published record can be saved and pasted in as:

   x x 39.000 41.000 49.000 37.000 x x 35.000 132.000 30.000 45.000 50.000 x 59.000 37.000 52.000 62.000 26.000 x x 52.000 x 65.000 x 36.000 52.000 34.000 x 66.000 56.000 x x x x x x x x x.

   After copying the above line into the WebQTL entry box, an SDP will be generated that will be used to compute the influence of QTL on the age-related thymic involution rate in BXD mice.
4. Optional choice: Enable Use of Trait Variance and entering of Trait Name.
5. Clicking on the "Next" button to go to the Trait Data and Analysis Form page.

6. To analyze data at the Trait Data Form Page, select appropriate options and one or more function buttons. A new window will open to display results and provide users access to a series of additional analysis tools.
   (a) Basic Statistics provides the number of informative strains, median, mean, variance, standard deviation, minimum, maximum, and range of values of the dataset entered. Furthermore, the Basic Statistic page contains a figure to be used for assessing whether or not the data set is normally distributed. The data are plotted against a theoretical normal distribution in such a way that the points should form an approximate straight line. Departures from this straight line indicate departures from normality.
   (b) Trait Correlations allows one to compare the entered BXD trait values with those of all other records in the databases, including at least 500 BXD published phenotypes, 300 BXD genotypes, BXD brain mRNA expression analyzed by the M430 or U74Av2 genechip, BXD cerebellum mRNA expression analyzed by the M430 genechip, and BXD hematopoietic stem cell mRNA expression analyzed by the U74Av2 genechip.

   As an example, the top 6 phenotypes were correlated with the BXD thymic involution slope. Interestingly, the correlation analysis indicates that the age-related thymic involution rate exhibited a strong negative correlation with our previous studies of age-related BXD trait values in susceptibility of aged BXD mice to develop breast tumor after implantation of these mice with the TS/A breast tumor cell line. The correlation plot of these two traits can be obtained by going to the correlation value –0.7923 shown in the WebQTL website. This type of analysis leads to a new hypothesis that the age-related loss of new thymic emigrant may be associated with a decline in peripheral CTL response to TS/A in BXD RI mice. We have indeed observed that these two traits exhibited a strong negative correlation ($R^2 = 0.62$, $p = 0.0025$).
7. Interval Mapping is used to localize QTL on the relevant chromosomes. It computes linkage maps for the entire genome and the consistency of linkage for a single trait can be assessed using permutation and bootstrap tests.
8. Marker Regression can be used to plot permutation results, list those markers linked to trait variation, and provide access to composite mapping functions. Loci in the BXD data set that have

associations with the entered trait data can then be shown in a Marker Regression Report table. The linkage map of the QTLs that exhibit association with the primary trait can be assessed by selecting the mapped chromosome and then going to Interval Mapping, located at the Trait Data and Analysis Form Page.

9. The Pair-Scan feature provides an analysis of epistatic interactions of two separate loci. The interaction is displayed by a pair-scan result for the entered trait. The left half of the plot highlights any epistatic interactions (corresponding to the column labeled "LRS Interact"). The lower right half provides a summary of LRS of the full model, representing cumulative effects of linear and nonlinear terms (column labeled "LRS Full"). This page also shows a table that list the LRS scores for the top 50 pairs of intervals (Interval 1 on the left and Interval 2 on the right). Pairs are sorted by the "LRS Full" column. Both intervals are defined by proximal and distal markers that flank the single best position. The pair interaction pattern can be viewed by going to the "Plot" located at the LRS Full column shown in the table.

Search of the Candidate Genes Located Near the Mapped QTL

This section provides a way to search for the candidate genes that are close to the mapped QTL. It must be emphasized that a search of candidate genes only provides a reference to judge if the mapped QTL region contains genes that can potentially influence the interested quantitative trait. This also serves as a reference for future detail gene cloning or a genetic manipulation study to confirm the mapped QTL. It, however, does not provide any guarantee that the candidate genes searched from the website are the genes influencing the mapped trait.

Genes located near the mapped QTL can be directly accessed from the UCSC Genome Browser and the Ensemble Genome Broswer websites by clicking at the corresponding genome section located on top of the interval mapping figure. Alternatively, the candidate genes can be searched using Mouse Genome Informatics website provided by the Jackson Laboratory. Again, we will use the mapped QTL associated with thymic involution on Chr 9 (*D9Mit243*) as an example for this demonstration.

1. Enter the website address: http://www.informatics.jax.org/.
2. Select *Genes and Markers* under the Search Menus.
3. Type in "D9Mit243" under *Symbol/Name*, select "9" under Chromosome, and leave all the other selections blank, then go to "Search."

4. The location of D9Mit243 on Chromosome 9 (61 cM) will be shown.
5. Go back to the Gene and Marker page.
6. Select "Gene" under Type, select "9" under Chromosome, select "50 and 70 cM" under cM position, leave all the other selection blank, then go to "Search."
7. All genes located between the selected chromosome regions will be shown.
8. Go back to the Searches of Gene and Marker page.
9. Select "QTL" under Type, select "9" under Chromosome, select "50 and 70 cM" under cM position, leave all the other selection blank, then go to "Search."
10. All previously reported QTL between the indicated region (50 to 70 cM on Chr 9) will be reported. In this case, the current mapped locus is indicated as the age related thymic involution 1 on this page.

Moving From a QTL to Gene(s) in Mice

Generally, but not necessarily, genes that exhibit polymorphisms between the parental strains, exhibit a known function that may influence the quantitative trait, and are expressed at the appropriate cell types, are more likely to be the candidate genes than the ones that do not have these characteristics. Another approach to help to determine if a gene located near the mapped QTL would have effects to influence the quantitative trait will be to use genetically engineered mice to determine if altering the expression of a candidate gene will alter the phenotype of interest.

However, it is possible that a quantitative trait is a combined effect of multiple genes located near the QTL. Thus, simply altering one gene may not necessarily provide a comprehensive link of the candidate genes with the quantitative trait, and in some cases, a false-positive result may even be obtained using the QTL analysis approach. Ideally, one should use additional BXD RI strains, other genetic background of RI strains, recombinant inbred cross (RIX), or F2 mice to confirm the mapped QTL before moving on to identify the genes that play an important role to influence the interested quantitative trait.

The detailed procedures that enable one to move from QTL to genes in mice are beyond the scope of this chapter. We will therefore only discuss some possible approaches that are currently used.

Fine mapping of the QTL

Mackay outlines approaches for moving from a QTL to a gene. The two important statistical considerations of QTL mapping are power and significance threshold. In cases in which the power is low, not all QTLs will be detected, leading to poor repeatability of results and an overestimation of the effects of QTLs that can be detected. Increasing the number of samples can increase the power. The power can also be increased by using the logistic curves that mea sure phenotype using multiple numbers of genetically fixed BXD RI strains at a given age and also by measuring the phenotype at carefully selected ages to provide higher power by fitting this curve with the best curve fit. Our data suggest that thymic involution in mice can be best fit with a negative exponential curve. However, high-resolution mapping can be overcome by adding more BXD RI mice per set. Jeremy Peirce, Lee Silver, and Robert Williams have generated and mapped an additional 40 BXD RI strains that have undergone advanced intercross and, therefore, have close to twofold map expansion (in G10 mice) as in the traditional BXD RI set. Members of the Complex Trait Consortium are initiating an eight-way cross to generate 512 RI lines for complex trait analysis. This should provide high genetic diversity and sub-cM mapping resolution. Strains with crossovers within the QTL can be further used to provide higher precision mapping and confirmation of a QTL.

Verification of QTL analysis using congenic mice

Following the linkage analysis of the BXD RI strains, congenic mice can be produced to isolate the effect of the genetic region of interest on different backgrounds to enable an analysis of the genetic effect of this interval. Quantitative traits in general are regulated by different loci/genes that contribute to a varying degree to the observed phenotype. Unpredictable interactions between unlinked genes that affect a phenotype may hamper the interpretation of data obtained in a regular F2 intercross. Therefore, a worthwhile intermediate step in the ultimate cloning of such quantitative genes is the construction of congenic mice. In this way, the effects of each quantitative locus can be studied independently. Because it is possible to map the genetic locus containing the desired traits within 3.75 cM using the BXD RI strains of mice, congenic mice will be selected using analysis of the different loci of interest in nearby DNA markers-the marker-assisted selection protocols (MASPs). This technique makes it possible to reduce the length of the differential chromosome segment more rapidly by screening back cross offspring for the occurrence of cross over using these DNA markers. Reciprocal congenic strains can be generated to allow the investigator to determine: (1) the effect of a B allele on D background; (2) the effect of a D allele on B background; and (3) Interaction of the mapped QTL with other background loci. Loci identified by the BXD RI linkage analysis can be transferred to both progenitor strains by a backcross breeding scheme. F1 mice can be backcrossed onto both C57BL/6J and DBA/2J mice. Offspring of each generation can then be genotyped and mice carrying the allele of interest can then be further backcrossed with the progenitor strain of the opposite genotype for four generations.

The method is essentially the same as that of speed congenics as reported by Lander and Schork to describe congenic strains developed in three to four backcrossed generations by marker-director breeding. Based on genotyping results, the mouse with the fewest chromosomes containing donor origin material can be bred. If there is more than one eligible male, the sum of donor original chromosomes eliminated is calculated and only males for which this sum is greatest remain eligible to father the next generation. This is superior to using the strategy that sums the length of the eliminated donor origin chromosomes and selects for a mouse in which the sum is greatest, but disregards the total number of donor origin chromosomes that are eliminated. Using this protocol, the numbers of fathers required for each generation and the numbers of markers required per generation results in the

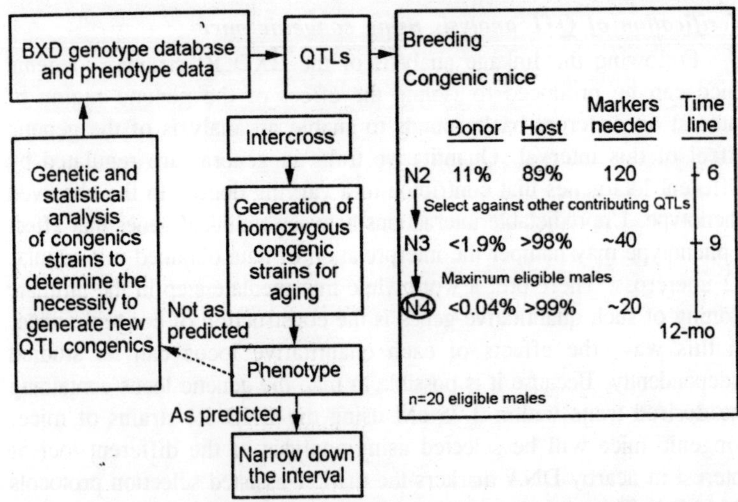

Fig. 12.2. Summary of the strategy to be used for generation of marker-assisted speed congenic mice.

production of the least donor genetic material by generation four (N4). Using this strategy, by N4 only 0.04% of donor genetic material remains and the mean number of chromosomes with retained donor genomic material is 0.78.

Notes

1. LRS provides a measure of the linkage between variation in the phenotype and genetic differences at a particular genetic locus. LRS values can be converted to LOD scores (logarithm of the odds ratio) by dividing by 4.61. The LRS itself is not a precise measurement of the probability of linkage, but in general for F2 crosses and RI strains, values above 15 will usually be worth attention for simple interval maps. The LOD ratio provides a measure of the linkage between variation in the phenotype and genetic differences at a particular genetic locus. The LOD is also roughly equivalent to the -log(P), where P is the probability of linkage (P = 0.00 1 = >3). LOD score analysis is equivalent to likelihood ratio testing, but instead of natural logarithms, logs to the base 10 are used. In the linkage analysis, the parameter of interest is the recombination fraction (θ) between marker and the locus that is compared. The null hypothesis represents no linkage between QTL and marker locus ($\theta = 0.5$), and the alternative hypothesis assumes linkage exists ($\theta < 0.5$). The LOD score function s then defined as:

$$\text{LOD}(\theta) = \log_{10}[\text{like}(\theta)/\text{like}(\theta=1/2)]$$

2. Measurement of secondary phenotypes is important as a second method to determine the accuracy of the primary quantitative trait of interest.
3. The method used for preparation of single cell suspension of thymocytes for determination of thymocyte count is essentially the same as that described by Bader.
4. One difficulty associated with the genetic analysis of thymic involution is that the initial thymus size may affect determination of the thymic involution rate. That is, although these two traits may be independent of each other, the initial size may confound the ability to determine the involution rate if one simply determines and compares the thymocyte count in young and old mice. Our study using BXD RI strains of mice clearly demonstrated this concept, because all strains and the parental B6 and D2 strains have varied initial thymocyte count and thymic involution rate, and the initial thymocyte count does not necessarily correlate with

the thymocyte count at older age. Therefore, in order to present thymic involution as a functional alteration with age, we have applied a mathematical model to generate a quantitative trait that can best be used to represent the change of thymus over time. This type of approach has been used to obtain the QTLs for growth- or other age-related traits.

5. Akaike's Information Criterion (AIC), introduced almost 30 yr ago by Akaike, is an information criterion for the identification of an optimal model from a class of competing models. Mathematically, AIC = ln (s_m^2) + 2m/T, where m is the number of parameters in the model, and s_m^2 is (in an AR(m) example) the estimated residual variance: s_m^2 = (sum of squared residuals for model m)/T. That is, the average squared residual for model m.
   The criterion may be minimized over choices of m to form a tradeoff between the fit of the model (which lowers the sum of squared residuals) and the model's complexity, which is measured by m. Thus an AR(m) model vs an AR(m+1) can be compared by this criterion for a given batch of data.
   An equivalent formulation is this one: AIC = T ln(RSS) + 2K where K is the number of regressors, T the number of observations, and RSS the residual sum of squares; minimize over K to pick K.
6. The negative exponential curve used as a quantitative trait in each of the BXD RI strains of mice reflects thymic involution from early adult to late adult age. Other factors that affect thymic size in young mice or very old mice may not follow this curve, and will not be represented in QTLs using this particular quantitative trait. The BXD thymic involution quantitative trait described in this chapter can be retrieved from the WebQTL website BXD Phenotype Database. The procedure to obtain this dataset is as follows:
   (a) Enter the GeneNetwork website address: http://www.genenetwork.org/home.html.
   (b) Select the most appropriate database for your particular query. In this example, we focus on the mouse BXD set.
   (c) Select "Mouse" under Choose Species, select "BXD" under Group, select "Phenotypes" under Type, select "BXD Published Phenotypes" under Database, enter "10231" under Term, then click the search button.
   (d) At the new page, click on RecordID/10231 - Age-related thymic involution slope (–Beta 1×10^4) by Hsu HC et al.

(e) This will lead to the opening of the Trait Data and Analysis Form for the BXD age-related thymic involution -β1 value quantitative trait page for the subsequent genetic linkage analysis using the analysis and mapping tools provided by GeneNetwork.

7. How should one handle thc quantitative trait data set when the SDP is not under a normal distribution pattern? Many statistical methods for QTL mapping, such as the EM algorithms of interval mapping and composite interval mapping, were developed based on the assumption that the phenotype is normally distributed. Thus, the first step in the analysis of most traits should usually be to review the data and its distribution. This is done by selecting the "Basic Statistics: function toward the top of the GeneNetwork Trait Data and Analysis Form." The goal is to determine whether the distribution is sufficiently close to normal to allow the use of standard parametric statistics. A standard procedure in QTL mapping is to transform the phenotype to resemble normality as closely as possible prior to analysis. A change in scale does not alter the information content of the original data. It simply changes the relationship of character values to one another. A variety of transformations can be used. The Box-Cox transformation provides a fairly general class of transformations for achieving normality. The Box-Cox transformation is

$$y_i^{(\rho)} = \begin{Bmatrix} (y_i^{\rho} - 1)/\rho & \rho \neq 0 \\ \log(y_i) & \rho = 0 \end{Bmatrix}$$

where y_i is the original phenotype, and ρ is a parameter. Box and Cox developed a method to find the ρ that gives the best fit to normality. This transformation includes the original data (ρ = 1) and the log transformation (ρ =0) as special cases.

It is also important to search for and handle any outlier cases. In the case of Record 10231, one will notice that BXD8 is a high outlier. It will have a disproportionately large effect on some types of analysis unless we use rank order statistics. For purposes of mapping, it will also have too much an effect unless we windsorize BXD8 ("windsorize," after the British monarch, King Henry the VIII of Windsor, who had the habit of chopping off the heads of his troublesome wives and colleagues). Our goal in windsorizing is to keep the data for BXD8, and to retain the advantage of parametric procedures. To do this, we truncate the value of BXD8 from 132 to 70. All of the values in the Trait

Data page can be modified temporarily by any GeneNetwork client. Once BXD8 has been windsorized to a value of 70, it is still the highest value, and the dataset can now be used for the linkage analysis.

8. The permutation test, which is taken directly from the work of Churchill and Doerge, is a method for establishing the significance of the likelihood ratio statistic generated by the interval mapping procedures. In this test, the trait values are randomly permuted among the progeny, destroying any relationship between the trait values and the genotypes of the marker loci. The regression model is fitted for the permuted data at all positions in the genome and the maximum likelihood ratio statistic is recorded. This procedure is repeated hundreds or thousands of times, giving a distribution of likelihood ratio values expected if there were no QTL linked to any of the marker loci.
9. The WebQTL implementation of the scan for 2-locus epistatic interactions is based on the DIRECT global optimization algorithm developed by Ljungberg, Holmgren, and Carlborg. In general, this method should not be used until the data set is quite large—preferably well over 50 cases or strains.
10. It is worthwhile to point out that simply moving the "D" type or "B" type QTLs on B6 background or D2 background, respectively, may not necessarily result in expression of the expected phenotype in the congenic mice. One consideration is that with respect to each trait, there are other QTLs that contribute to each of the phenotypes. For example, complementation by these QTLs in the B6 background (and thus of B6 origin) may modify the overall expression of the D2-derived QTLs in the congenics and it is not possible to predict whether they will increase or decrease variation seen between pure-strain B6 and DBA mice. Generation of reciprocal congenics and testing the model may allow one to determine whether the QTL region has an effect on age related changes of interest and whether there is an interaction of this QTL with background genotype, thus reduce the chance to observe no effects of the QTL in congenic mice.

Genetic Polymorphisms

Studies in model organisms using mutant and transgenic animals have shown that life span is influenced by genes that confer metabolic efficiency, promote vitality and resistance to stress. Yet, these studies may be of only limited relevance for our understanding of the genetic causes of variation in life span in natural populations, which are important for understanding its maintenance and evolution. Longevity is a complex quantitative trait and segregating genetic variation for longevity has been shown to exist in natural populations, with heritabilities between 10 and 30%. Therefore, one way to address questions related to its maintenance and evolution in natural populations is to use a quantitative genetic approach. This allows the identification of loci that contribute to natural variation in life span and the estimation of the relative effects of these loci across genetic backgrounds and different environments.

There are two steps in the quantitative genetic approach to identifying naturally occurring genetic variation affecting longevity. The first step involves mapping the *quantitative trait genes* (QTGs) responsible for producing variation in life span between two strains. This is followed by *linkage disequilibrium* (LD) mapping in a large sample of alleles collected from a natural population to determine the molecular polymorphisms (*quantitative trait nucleotides* [QTNs]) within candidate genes that may be responsible for the variation in life span. Identification of candidate genes starts with recombination mapping using F2 populations, informative backcrosses, or recombinant inbred lines derived from two progenitor strains showing different life spans. This allows the identification and enumeration of the chromosomal regions containing loci responsible for producing variation in life span (quantitative trait loci [QTLs]) between the two progenitor strains.

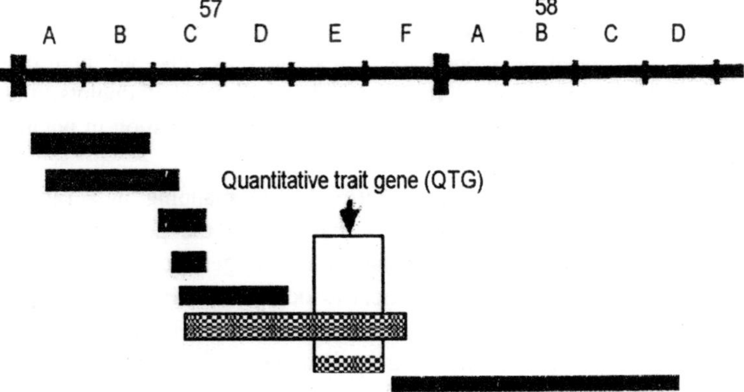

Fig. 12.3. Representation of quantitative deficiency mapping of a D. melanogaster chromosomal region (57A-58D) containing a hypothetical QTG affecting variation in life span.

However, identification of candidate loci from the results of QTL mapping alone is problematic in most organisms because the confidence limits on QTL boundaries are often quite large, and usually encompass anywhere from tens to hundreds of genes.

In *Drosophila*, identification of candidates is straightforward because of the development of quantitative complementation tests to (1) deficiencies or deletions on one of the chromosomes to refine the QTL region of interest and (2) null mutant alleles of candidate genes to test the candidacy of the QTG. Deficiency stocks uncovering almost 90% of the whole *Drosophila* genome are commercially available from the main stock centers in Bloomington, Indiana and in Umea, Sweden. The size and location of the available deficiency stocks are different and the boundaries of the deleted sections are known. After obtaining a set of deficiency stocks that, together, completely span the QTL region of interest, one can test the effects of each small deletion on the trait.

Sequential tests of the effects of each small deletion within the originally identified larger QTLs can reduce the region of interest to a manageable size so that candidate genes can be identified. If a mutant stock of a specific gene localized in the QTL boundaries is available, a complementation test similar to the deficiency test is carried out to identify the positional candidate gene. A possible problem that one could face is that no mutations of the candidate gene are available in the refined chromosomal region. Yet, the ongoing effort by the Berkeley *Drosophila* Genome Project to develop precision mapped *P*-element mutations in all known *Drosophila* genes will soon greatly reduce this limitation. Additional methods and resources have also recently been developed for researchers to generate their own custom deletions of target regions with predictable endpoints, thus creating their own stocks for testing the effects of variation in putative candidate genes.

Once the QTG has been identified, LD mapping in a random mating population can be used to characterize the genetic basis of natural variation in life span at the molecular genetic level. The LD mapping method tests for a statistical association between polymorphic markers within the QTG and phenotypic variation in life span. If a polymorphic site at a candidate gene is associated with differences in the quantitative trait phenotype, that site either is the QTN or is in LD with the QTN. As LD denotes the nonrandom association of alleles at different loci in gametes and it does not provide direct proof that the allelic variation in the candidate polymorphism(s) under study is the causal agent of phenotypic variation. This is because LD can be generated by recent population admixture, selection, and inbreeding, and so the true causal polymorphism might be linked to the putative causal site identified in the LD study.

As LD between polymorphic loci is expected to decay through time as a function of the recombination fraction between them, over long periods of time only very closed linked polymorphisms will remain in LD. Therefore, the use of a dense map of polymorphic markers within the QTG can be used to track down the QTN. Formal confirmation of the phenotypic effect of identified polymorphisms requires additional genetic manipulations (e.g., homologous recombination) to examine the influence of alternative alleles of the candidate gene on the phenotype in different genetic backgrounds. However, the mapping procedures outlined in this chapter have proven useful for identifying the candidate genes on which to focus our efforts.

Extensive work has been conducted in the laboratory of Trudy Mackay, where we both trained, to unravel the genetic basis of naturally occurring variation for longevity in *D. melanogaster* using recombination mapping and the quantitative complementation approach outlined in this chapter. Based on those endeavors, *Dopa decarboxylase* (*Ddc*), a gene encoding the enzyme DOPA decarboxylase which catalyzes the final step in the synthesis of the bioamines dopamine and serotonin, was identified as a candidate gene affecting *D. melanogaster* longevity. Deficiency mapping was first used to narrow one QTL region to a 50-kb interval that contained bioamine biosynthesis pathway genes. Then, a complementation test for three *Ddc* alleles was performed to show that *Ddc* is a positional candidate gene for life span in these strains. Finally, the effects of naturally occurring polymorphism in the *Ddc* gene on the life span of virgin flies were tested using chromosome 2 substitution lines (C2SLs) established from the natural population of *D. melanogaster* in Raleigh, NC.

C2SLs were established from 173 isofemale lines created from the *D. melanogaster* females caught in the wild. They were constructed by extracting a single second chromosome from each of the 173 isofemale lines and substituting it into a common genetic background, the Samarkand (*Sam*) strain, using a standard *Drosophila* crossing. *Ddc* was completely sequenced from 14 of the C2SLs that varied in longevity to identify single-nucleotide polymorphisms (SNPs) to use as genetic markers to screen the complete set of 173 alleles. The C2SLs were genotyped for 36 genotypic markers. Three common SNPs, T420C, C1685A, and T2738G, which were in strong LD, were associated with naturally occurring variation in longevity. The extensive LD throughout the 5.5-kb region, including the *Ddc* transcription unit, hinders our ability to infer the QTN causing the observed variation in life span. Further studies will necessary to prove that molecular polymorphisms at *Ddc* cause variation in life span.

However, it is important that the overall result of our study—multiple changes within a single locus are correlated with naturally occurring phenotypic variation—is supported very strongly by similar research on alcohol dehydrogenase (ADH) activity and bristle number variation in *D melanogaster*. If multiple intralocus effects and interaction turn out to be the rule for QTL, studies using *D. melanogaster* as a model system will play an important rôle in identifying the genetic variants responsible for naturally occurring variation in life span. The identification of these variants will provide a better understanding of the mechanisms by which variation for longevity is maintained in natural populations.

One note of caution regarding the interpretation of life span studies carried out as described below. It is important to recognize that life span may be influenced by genetic loci that do not necessarily influence the rate of aging or senescence (defined as the physiological deterioration with age).

Genetically based variation in any phenotypic trait that affects the probability of survival, independent of age (e.g., copulation frequency, reproductive output) would result in differences in life span. Thus, many genes that influence variation in life span may have no influence on determining rates of senescence. Once candidate genes affecting life span are identified, follow-up experiments would need to be done to determine if allelic variation at these loci also caused variation in age-specific rates of decline in fitness or instead produced variation in age-independent factors that increased the probability of death.

Materials

Quantitative complementation tests to deficiencies and mutations

1. The two progenitor strains from which the segregating population used to map QTL was originated.
2. Drosophila stocks.
3. SAS software.

Linkage disequilibrium mapping

Sequencing

1. Gentra DNA isolation kits.
2. 0.2-mL, 96-well PCR plates.
3. PCR reagents: 10X PCR buffer; 2 m*M* dNTPs; 25 m*M* $MgCl_2$; 10 μ*M* forward and reverse primers; 5 U/μL of Taq DNA polymerase.
4. Standard agarose.
5. 10X TAE buffer: 96.8 of Trizma Base, 40 mL of 0.5 *M* EDTA pH 8.0, 22.84 mL of glacial acetic acid; add water up to 400 mL (pH 8.0–8.1).
6. 10 mg/mL Ethidium bromide.
7. DNA ladder.
8. QIAquick 96 PCR Purification kit.
9. ABI PRISM BigDye Terminator Cycle Sequencing Ready Reaction mix (PE/Applied Biosystems, Foster City, CA).
10. Nonskirted 0.2 mL 96 PCR plate.
11. 10 μ*M* of each primer.
12. 70% ethanol.
13. Formamide.
14. DNA alignment software.

Genotyping: pyrosequencing

1. PCR reagents: 1X PCR buffer; 2.5 m*M* of dNTP; 25 m*M* $MgCl_2$; 10 μ*M* downstream and upstream primers; 5 U/μL AmpliTaqGold polymerase.
2. 0.2-mL, 96-well PCR plates.
3. Plate sealers.
4. Pyrosequencing Instrument, PSQ96.
5. Thermomixer R.
6. Streptavidin-coated Sepharose beads (Pyrosequencing AB).
7. Binding buffer pH 7.6 (10 m*M*Tris-HCl, 2 NaCl, 1 m*M*EDTA, 0,1% Tween 20).
8. Denaturation solution (0.2 NaOH).
9. Washing buffer pH 7.6 (10 m*M* Tris-Acetate).
10. Annealing buffer pH 7.6 (20 m*M* Tris-Acetate, 2 m*M* $MgAc_2$).
11. 3 μ*M* Pyrosequencing primer.
12. 96-well filter plate.
13. 96-well PSQ96 plates.

14. PSQ96 reagent kit.
15. SQA cartridge.

Methods

Quantitative complementation tests to deficiencies and mutations

Longivity phenotypes

1. Cross 10 virgin females of each parental line to eight males of each deficiency stock or mutation stock.
2. Rear flies on standard cornmeal/agar/sugar food *with* live yeast in an incubator at 25°C, 60–75% relative humidity and a 12-h dark-light cycle.
3. Remove adults after four days of mating and egg-laying.
4. On day 12–13, randomly collect virgin flies of the same genotype and sex and place them in a vial containing approx 5 mL of standard cornmeal/agar/sugar food *without live* yeast.
5. Set up five replicate vials for each genotype and sex each containing 5 virgin flies, for a total of 25 males and 25 females per genotype.
6. Transfer flies to fresh medium every 5 d, and record the number of dead flies in each vial daily.

Data analyses

1. Estimate longevity as the number of days a fly lived, from the day of eclosion to the day prior to registration of death.
2. Using GLM SAS procedure, analyze the data for each sex separately by a two-way mixed model analysis of variance (ANOVA):

$$y = \mu + L + G + LXG + R(L \times G) + E$$

   where p is the overall mean, L is the fixed main effect of line (P_1 or P_2), G is the fixed main effect of genotype (*Def* or *Bal*), and E is the error term. The term $R(L \times G)$ indicating replicate vials nested within line and genotype is included in the analysis to remove vial effects. A significant line-by-genotype ($L \times G$) interaction term is indicative of a failure to complement. There is also an additional criterion that is typically used in interpreting the results from this type of analysis. The difference in the average phenotypes lines of each line over the defi-ciency ($P_1/Def P_2/Def$) must be greater than the average phenotypes when each parental chromosome is with the balancer ($P_1/Bal - P_2/Bal$). Greater differences between the parental chromosomes when they are with the balancer indicate epistatic interactions between the QTL in the parental lines with genes on the balancer chromosome.

Linkage disequilibrium mapping

Longevity phenotypes

1. Cross 10 females of each C2SL1 to 8 males of the same line.
2. Rear flies on standard cornmeal/agar/sugar food *with* live yeast in an incubator at 25°C, 60–75% relative humidity and a 12-h dark–light cycle.
3. Remove adults after four days of mating and egg-laying.
4. On day 12–13, randomly collect virgin flies of the same genotype and sex and place them in a vial containing approx 5 mL of standard cornmeal/agar/sugar food *without live* yeast.
5. Set up five replicate vials for each genotype and sex each containing 6 virgin flies, for a total of 30 males and 30 females per genotype.
6. Transfer flies to fresh medium every 5 d, and record the number of dead flies in each vial daily.

DNA isolation

Gentra DNA isolation kits are available to isolate DNA from a single fly up to 700 flies. DNA isolation is performed according to the instructions provided with the PUREGENE DNA isolation kit. The expected yield ranges from 0.3 to 700 pg of DNA depending on the number of flies used for the isolation. DNA is quantified spectophotometrically.

DNA amplification and sequencing

PCR primers are designed using the published sequence of *D. melanogaster* to amplify overlapping segments of the gene region. The PCR products are sequenced directly from both strands using a set of internal primers, designed at interval of approx 200 bp, and ABI big dye terminator chemistry.

1. Prepare a PCR master mix for 100 reactions adding the following: 1.24 mL of purified water; 250 μL of 10X PCR buffer; 150 μL of 2.5 m*M* dNTPs; 150 μL of 25 m*M* $MgCl_2$; 100 μL of 10 μ*M* forward primer; 100 μL of 10 μ*M* reverse primer; 10 μL of *Taq* DNA polymerase.
2. Place 5 μL of each DNA (200 ng) in each well of a 0.2-mL 96-well PCR plate and dispense 20 μL of the master mix. Include a negative control containing 5 μL of water instead of DNA.
3. Run a standard PCR program characterized by an initial denaturation step at 94°C for 5 min and followed by a three-step profile with a denaturation at 94°C for 30 s, primer annealing at

optimal anneal temperature for 30 s, and extension at 72°C for 45s, for a total of 35 cycles.

4. Control the PCR yield running each PCR product on a 2% agarose gel in 1X TAE buffer containing 1 μg of ethidium bromide per mL. Add 10 μL of PCR products to each well (5 μL of PCR product and 5 mL of 2X Ficoll loading dye), and add in one well 10 μL of a size ladder to confirm the correct size of the PCR products. Run at 100 mA until the xylene cyanol dye is approx 2 cm from the bottom of the gel. Place the gel on the UV transilluminator to visualize the PCR products.
5. Purify products using a QIAquick 96 PCR Purification kit according to the instructions provided with the kit.
6. Prepare a master mix for 100 reactions containing: 400 μL of BigDye Terminator Mix and 100 μL of 1.6 pmol/mL sequencing primer.
7. Place 5 μL of each PCR product in each well of a nonskirted 0.2-mL, 96-well PCR plate and dispense 5 μl of the master mix. Include a negative control containing 5 μL of water instead of DNA.
8. Perform sequencing reaction in PCR machine using standard cycling method: 96°C for 10 s, 50°C for 5 s, and 60°C for 4 min for a total of 35 cycles followed by 4°C hold.
9. To purify reactions add cold 70% ethanol to each sample and mix.
10. Precipitate at room temperature for 15 min exactly.
11. Spin plates at 4000 rpm for 30 min.
12. Discard ethanol and dry pellets, cover plate and store at −20°C until needed, or add 10 μL formamide loading buffer and run on the sequencer immediately.
13. Align sequences using DNA alignment software and identify SNPs.

Genotyping assay: Pyrosequencing

PCR amplification

1. Prepare a PCR master mix for 100 reactions. For a 50-μL PCR reaction, the following reagent concentrations must be used: 1X PCR buffer; 0.2 m*M* of each dNTP; 1.5–3.0 m*M* $MgCl_2$, 0.2 m*M* downstream and upstream primers; 1 U AmpliTaqGold polymerase. Either forward or reverse primer is biotinylated.
2. Place 5 μl of each DNA (250 ng) in each well of a 0.2-mL, 96-well PCR plate and dispense 45 μL of the master mix. Include a negative control containing 5 μL of water instead of DNA.

3. PCR reaction starts with a 95°C denaturation for 5 min. This is followed by a 50-cycle thermal cycling. Each cycle is programmed to include 15 s denaturation at 95°C, 30 s annealing at appropriate temperature, and a 15–30 s final extention at 72°C.
4. Check on 2% agarose gel the size and quantity of your PCR product.
5. Cover plate and store at 4°C until needed.

Sample preparation

1. Add 50 μL of binding buffer to the 50 μL of PCR product to immobilize biotinylated PCR products on streptavidin-coated Sepharose HP beads.
2. Add 4 μL of streptavidin-coated Sepharose beads and mix at 1400 rpm in a Thermomixer R at room temperature for 10 min.
3. Transfer the streptavidin-coated Sepharose bead and PCR mixture to a 96-well filter plate and remove excess fluid by vacuum filtration.
4. Denaturate the biotinylated DNA attached to the streptavidin-coated Sepharose beads in 50 μL of denaturation solution for 1 min, allowing the nonbiotinylated strand to be removed.
5. Remove the denaturation buffer by vacuum and wash DNA twice with 150 μL of washing buffer.
6. Resuspend DNA in 50 μL of annealing buffer.

Pyrosequencing reaction

1. Mix the DNA well and transfer 40 μL of it to a 96-well PSQ96 plate.
2. Add the appropriate sequencing primer in a volume of 5 μL using a 3 μ*M* stock solution, resulting in 45 μL reaction volume.
3. Anneal the sequencing primer on a heat plate set for 80°C for 2 min.
4. Allow the samples to cool for 5 min at room temperature to facilitate primer annealing to template.
5. Once samples have cooled down, place the plate on the Pyrosequencer and add the PSQ96 reagents to the SQA cartridge following the order indicated by the manufacturer.
6. Pyrosequencing data output is quantified using Peak Height Determination Software v1.1.

Data analysis

1. Estimate longevity as the number of days a fly lived, from the day of eclosion to the day prior to registration of death.

2. Assess associations between molecular polymorphisms and longevity by one-way factorial ANOVA of line means, according to the model

$$y = \mu + M + S + M{\times}S + E,$$

where *M* and *S* denote the fixed effects of molecular marker and sex, respectively. A statistically significant *M* indicates an association between variation at the molecular marker and variation in life span.

Notes

1. Deficiency and mutant stocks are available from two main *Drosophila* stock centers: Bloomington Stock Center and the Szeged *Drosophila* Stock Center.
2. The number of parental females used to set up the cross is simply the number we have used in our assays. Increasing or decreasing the number of females will increase the density of larvae in a vial which can in turn affect life span. The important consideration here is to have comparable larval densities in the vials from which the assayed individuals are taken.
3. The number of replicate individuals that should be used in any experiment will vary depending on the amount of variation within and between genotypes and the minimal detectable difference in life span between genotypes that one wishes to detect. Data from pilot projects can be used to estimate the expected variation within and between treatments and power calculations made to determine the sample size required to detect life span differences of a given magnitude.
4. In the BigDye Terminator Cycle Sequencing Ready Reaction mix, the dye terminators, deoxynucleoside triphosphates, AmpliTaq DNA Polymerase, magnesium chloride, and buffer are premixed into a single tube and are ready to use.
5. One of the primers is biotin labeled for immobilization to Sepharose beads. The other is unlabeled. As free biotin may compete with the biotinylated PCR product for binding to streptavidin, thereby lowering the signal level, it is recommended that one purify the biotinylated primer by HPLC or equivalent procedure.
6. Biotinylated PCR primers are particularly sensitive to storage. Dilute primers from the stocks and aliquot them. Keep stock primer and aliquots of the diluted primers at –20°C, *not* at 4°C.
7. Design of sequencing primers for Pyrosequencing follows the same

criteria as for the PCR primers, except that the T*a* of this primer may be lowered. Primers with T*a* around 50°C work well in most cases. The interpretation of the SNP will sometimes be difficult if one or both of the polymorphic bases will form a homopolymer with adjacent bases. Therefore, consider designing your sequencing primer to avoid such a situation by placing the 3'-end to overlap the polymeric stretch. However, if this is unavoidable, up to 5 C, G, T, or 3 A polymers are acceptable.

8. Due to the size of the experiment, life span assays may need to be conducted in separate temporal blocks.
9. To minimize the contribution of polymerase error to sequence variation, it is recommended that one set up for each line several 50-μL PCR reactions from each primer pair and pool them together before purification with the QIAquick 96 PCR Purification kit.
10. Make sure that no liquid remains in the tubes, or unincorporated terminators will be carried over.
11. Select PCR primers to amplify a fragment that is $<$300 bp. The PCR primers should (a) be at 18–25 bp in length, (b) be more GC rich in the 5'-end and less in the 3'-end, and (c) not form heavy hairpin loops or dimmers with themselves or the other primer. As the precision of the Pyrosequencing reaction depends on the specificity of the PCR product, the choice of the primers is an important matter; therefore, for a more accurate selection of each primer set, it is strongly recommended that one use a software package that automates their design.
12. Make sure that the PCR product is clear, and that no excess primers, primer dimers, or other nonspecific products are present.

13

AGING AND LIFE SPAN IN MICE

There is the growing scientific and public interest to the development of new anti-aging drugs. Accordingly, there is the need to create standard guidelines for testing such drugs and for evaluation of life extension potential as well as other late effects (like carcinogenic and cancer preventive effects of chemicals). One of the approaches was used by International Programme for Chemical Safety. Some principles of a program for testing biological interventions to promote healthy aging were discussed in US National Institute on Aging.

Guidelines for testing should include such significant points as animal models, regime of testing and biomarkers/endpoints. Mammals are most appropriate animal models, because their biology is sufficiently homologous to that in humans. Mice are the most appropriate models in terms of husbandry, costs, and length of life. One of the most significant problems is the choice of mouse strain. To have a strain with genetic diversities and spectra of pathologies close to those of human population, application of four-way crossed mice is proposed by some authors. Another way to solve this problem is to use different strains in one study. That approach was used in our studies where we combined outbreed (SHR or NMRI) mice and inbred CBA mice as well as senescent-accelerated SAMP-1 mice and transgenic HER-2/neu mice.

Testing regimen should be nontoxic, simple, not stressful, and appropriate for human application. During the study, a number of biomarkers of aging could be included into the battery of tests. Among those, slit lamp exam for cataract, cognitive function assessment (maze learning), biomarkers of oxidative stress (diene conjugates, Schiff bases, malone dialdehyde, SOD, catalase, glutathion peroxidase, and so on), hormones (glucocorticoids, sex hormones, gonadotropins, prolactin, insulin, IGF-1, growth hormone, leptin, thyroxin, triiodthyronine), metabolic parameters (blood glucose, cholesterol, β-lipoproteins, triglycerides, free fatty acids), gene expression profiling with microarray technology, and so on could be performed. However, some methods are very costly and require a larger number of animals in the groups. For most of our studies, aside from survival parameters, we registered body weight, body temperature, food and water consumption, physical activity and muscular strength, and estimated estrous function. To evaluate late effects of drugs, we have also conducted pathomorphological examination, including tumor diagnostics. For general evaluation of the effect of drugs on aging, life span, and carcinogenesis, adequate mathematical and statistical models were used.

Materials

Animals

1. Strain choice. For strain selection, we recommend using two or more strains in one study, combining outbreed and inbreed strains. Additionally, for special mechanistic studies, genetically modified mice could be used.
2. The strain(s) must be well characterized in terms of their genetics, survival and pathologies. The most important aspect is a long-term experience in the use of selected strain in a laboratory where a study will be done.
3. Mice of both sexes could be used.

Housing Requirements

1. The animals should be kept in standard conditions of temperature and humidity.
2. Female mice should be kept in equal number per cage (5, 7, or 10) both in the control and experimental groups. Males should be kept single at a cage due to territorial fights.
3. Illumination should be standard as 12 h light/ 12 h darkness.
4. Animals should receive tap water and food (standard pellet) *ad libitum*.

Equipment and Reagents

1. Electronic balance (for estimation of body weight and food consumption).
2. Bottles with marked volume (for water consumption).

3. Plastic chamber 30 × 21 × 9 cm (for physical activity).
4. Electronic thermometer (for body temperature) is used.

METHODS

Age at Start of Treatment

Long-term studies should be started at the age of 2 mo, just after maturity of the females. For additional mechanistic studies, experiments could be started later.

Randomization in Groups

1. Just before the start of an experiment, animals should be randomly divided into control and experimental groups.
2. The number of animals should be sufficient for further statistical analysis (usually 30–50 mice per group).
3. All mice should be individually marked.
4. Control mice could be left intact and/or could be treated by the solvent using the same doses and regimen as in experimental groups.

Route, Dosage, and Regimen of Treatment

1. The best route of administration of a compound is with drinking water or supplementation to food. Some compounds that are low-soluble or nonsoluble in water might be dissolved in a small amount of ethanol and then dissolved for proper concentrations. For exact dosage of low-soluble compounds, gastric intubation could be used.
2. It is acceptable for parenteral route, if tested compounds are administered by subcutaneous injections (0.1 mL) of buffered solution in flank of the body. Intraperitoneal or intravenous injections are more stressful and risky for infection.
3. Testing two or more drug doses is recommended, including the dose proposed for usage in humans. Doses should be not higher than a *maximum tolerated dose* (MTD).
4. In a case of administration of a drug with drinking water or food, a treatment 5 or 7 d a week could be recommended. Gastric intubations are performed three to five times a week, parenteral injections are usually performed once a week.
5. The best regimen is long-term continuous exposure until natural death of the animals. However, in some cases, intermittent exposure could be used.
6. During the study, the animals should be under daily observation, with regular registration of the amount of drinking water and

consumed food, body weight and temperature, estimation of physical strength and activity, and estrous function.

7. The animals should be observed until their natural deaths, and sacrificed only when moribund. The date of each death should be registered and survival time should be estimated.

Food Consumption

This parameter should be monitored once a month during the all period of observation until a natural death of last animals in a group. Ten grams of a standard lab chow per mouse should be placed to cleaned food container in a cage, and the amount of nonconsumed food (in a food container and on the bottom of cage) should be weighted 24 h later; calculations of grams per mouse per day values should be performed.

Water Consumption

Water consumption should be also monitored monthly. A bottle with a drinking tap water should content 10 mL water per mouse and a volume of water drinked consumed in 24 h should be expressed in milliliters per mouse per day.

Body Weight

Individual body weight of mice should be measured monthly. We recommend measuring body weight at the same time of a day.

Body Temperature

Once every 3 mo, simultaneously with weighing, rectal body temperature should be measured with electronic thermometer.

The mice will be fixed in usual position sitting on grill with a tail turned up. The electrode of thermometer should be immersed in glycerol and introduced into rectum in a distance about 1.0 cm. Initially, rectal body temperature could decrease as a result of stress and angiospasm. Thus, it is necessary to wait until the mouse will be quiet and body temperature stabilized.

Estrous Function

1. Once every 3 mo, vaginal smears, taken daily for 3 wk from the animals, should be cytologically examined to estimate the phases of their estrous functions.
2. The thin cotton-ended sticks should be moistened with sterile water and smears should be taken and displayed at the preliminary marked glass for histological slides.
3. Cytological evaluation of the estrous phases should be performed 1–2 h after collection. It is possible to investigate nonstained

vaginal smears using a drop of a condenser of a microscope under ×80–100 magnification. The record of daily vaginal smears examination for each individually marked animal should be correctly performed and proestrus, estrus, metaestrus, and diestrus phases should be registered.

4. The following parameters of estrous function should be evaluated:
   (a) A length of each estrous cycle from the first estrous phase registered to each next first estrus day and then a mean length of the cycle in a group.
   (b) A ratio of phases of estrous cycle.
   (c) A rate of estrous cycles of various length (%). Usually we subdivide it as <5 days, 5–7 d, and >7 d.
   (d) Fraction of mice with regular cycles (%).
   (e) Fraction of mice with irregular cycles (%).

Method of Estimating Physical Activity of Mice in the "Open Field" Test

1. The mice should be tested at the age of 6, 9, 12, and 18 months in the daytime from 10 AM to 5 PM.
2. Animals of each group should be placed one by one in a plastic chamber measuring 30 × 21 × 9 cm, at the bottom of which squares (5 × 5 cm) were drawn: 5 squares in length and 4 squares in breadth.
3. Each mouse should be observed moving in the cage, and its locomotion parameters were estimated: (1) the number of crossed squares in the field (a square was considered crossed if the animal stepped over its border at least with two paws); (2) the number of vertical sets (when the animal rose to its hind paws); and (3) the duration of grooming reaction of muzzle, body, and genitalia. As a way to exclude the possibility of smell-associated orientation reaction, the chamber floor should be wiped with a wet cloth after each animal.

Method of Studying Muscular Strength and Physical Fatigability

1. The mice should be tested at the age of 6, 9, 12, and 18 months in the daytime from 10 AM to 5 PM.
2. After weighing, the mice are suspended on a string stretched to an altitude of 80 cm, so that they would hang by the string, clutching at it with their front paws.
3. The time until the moment of their fatigue and fall is registered in seconds.

4. In 20 min, the mice are suspended again and the time for which they managed to hold on is measured.
5. A discrepancy between these two indices is regarded as a parameter of physical restoration. It could be additionally estimated in relation to body weight.

Pathomorphological Examination

1. All the animals that died or were sacrificed when moribund, should be autopsied. At autopsy, their skin and all internal organs should be examined.
2. Revealed neoplasia should be classified according to the recommendations of the International Agency of Research on Cancer as *fatal* (i.e., those, that directly caused the death of the animal) or *incidental* (for the cases in which the animal died of a different cause).
3. Main internal organs, all tumors, as well as other lesions, should be excised and fixed in 10% neutral formalin. After routine histological processing, the tissues should be embedded in paraffin.
4. Thin, 5- to 7-μm histological sections should be stained with hematoxylin-eosin and microscopically examined; regarding the experimental group to which the mice belonged, this was a blind process.
5. Tumors should be classified in accordance with IARC recommendations.

Statistics

1. Experimental results should be statistically processed by the methods of variation statistics. The significance of the discrepancies should be defined according to Student's *t*-criterion, Fischer's exact method, a chi-square analysis, and a non- parametric criterion of Wilcoxon-Mann-Whitney.
2. For discrepancies in neoplasm incidence to be estimated, an IARC method of combined contingency tables calculated individually for the fatal and incidental tumors as well as a prevalence analysis should be applied.
3. For survival and risk analysis, Cox's method is the most useful. To test two groups survival equality, Taron's life table test can be used. All reported values for survival tests should be two-sided.

Mathematical Models and Estimations

1. The mathematical model could be used to describe survival under the treatment. The model is the traditional Gompertz model with survival function,

$$S(x) = \exp\left\{-\frac{\beta}{\alpha}[\exp(\alpha x) - 1]\right\}$$

where parameters α and β are associated with the aging rate and the initial mortality rate, respectively. Parameter α is often characterized by the value of mortality rate doubling time (MRDT), calculated as $\ln(2)/\alpha$.

2. Parameters for the model should be estimated from empirical data by use of the maximum likelihood method implemented in the Gauss statistical system.
3. Confidence intervals for the aging rate parameter estimates should be calculated by profiling the log-likelihood function.

Notes

1. Critical review of available data on the effect of life extension drugs has shown that from the point of view of the current guidelines a lot of studies are invalid. Tested drugs were often given to a small number of animals (10 to 20); the treatments started at the old age of animals, where a lot of more weak animals would die and more robust would survive; the observation stopped at the age of 50% mortality or at some other voluntary time, but not at the natural death of last survivor; the autopsy and correct pathomorphological examination sometimes were not performed; the body weight gain and food consumption were not monitored, and so on.
2. The special problem is the selection of mouse strain. For testing pharmacological drugs, genetic and other characteristics should be analyzed. For example, C57BL/6j mice are most common animal model in life span extension studies. However, because of genetically dependent deficiency in melatonin production by pineal gland in C57BL/6j mice this strain could not be adequate model. Some serious arguments support the use of hybrid or outbreed mice.
3. Spontaneous and induced genetic modifications, homozygous null mutations, knockout and transgenic mammalian animals were also introduced into experimental gerontology. It is worth noting that although the effects of some genetic manipulations are not expressed except at specific times in the animal's life because these genetic manipulations are in effect during embryonic development as well as throughout adult life, there are a significant limitations in the interpretations of the data obtained. Some of them were discussed

recently. On the one hand, the elimination of an activity or a pathway can lead to erroneous conclusion about the function of a gene because resulting compensation can markedly alter the physiology of the animal. On the other hand, overexpression of a transgene may readily yield no effect on life span or on any other aging parameter of the animal for that matter. Jazwinski have noted that overexpression of a transgene will likely create interactions with other genes and with the environment, both external and internal. Nevertheless, we have used mutant and transgenic mice for study of effect of some drugs on longevity.

4. Sex of mice is very important. We prefer to use female mice, which are not so aggressive as males. Another advantage of females is a possibility of a monitoring of estrous function and wider spectrum of tumors including than that in males.
5. In general, a fairly wide range of husbandry conditions yields similar life span results. Different frequency of cage changes (1, 2, or 3 wk) and different cage ventilation rates have not shown any differences in health status. Fairly dirty, unsanitary conditions had no significant effect on the life span. It seems also that animals raised in *specific pathogenic free* (SPF) conditions do not show a significant improvement in survival curves. While some reports did show an improved life span from germ free husbandry, it was, however, shown that germ-free animals had a lower caloric intake with that being the likely cause of life span extension.
6. The food quality and control are important issue. To know the content in lipids, kind of animal proteins, fibers, and relevant oligoelements may be important because of the great relevance of food quality used in animal facility, particularly when substances which may interfere with the nutritional status have to be studied. In our experiments, we routinely use the granular diet produced by Agricultural Company "Volosovo". This diet contains natural ingredients: wheat (30%), corn (22.7%), powdered milk (4.3%), soya (12%), meat meal (20%), yeast (5%), fat (3%), melassa (2%), and premix (1%). The total amount of the proteins in the food was 25.66%; fat, 7.0%; methionine, 0.7%; lysine, 1.44%; calcium, 2.04%; phosphor, 1.38%; natrium chloride, 0.37%; vitamins (mg/kg): A, 5000 i.e.; D, 500 i.e.; B_1, 1 mg; B_2, 2.5 mg; B_3, 2.5 mg; B_4, 120 mg; B_5, 12 mg; B_6, 1.5 mg; E, 50 mg; K, 10 mg; B_{12}, 0.03 mg; H, 02 mg; Cu, 8 mg; Fe, 60 mg; Co, 2.4 mg; Mg, 6 mg; Zn, 5 mg; iodine, 1.5 mg/total calories was

305 kilocalories per 100 g. In general, this diet is similar to NIH-07 rat and mouse ration.

7. An experiment must be initiated at the age after which growth has ceased, so that any life extension due to growth retardation is not a potential artifact. In some reports, a treatment have been started at 12 or 18 mo. We believe that use of such design as basic is not a correct for evaluation of premature aging preventive potential of the compound tested.
8. For correct choice of the route of drug administration, the data on solubility and stability of a compound in the water or in food are very important. If compound given with drinking water is sensible to illumination, bottles of dark glass should be used.
9. Concerning regimen of treatment, intermittent exposure could be only recom ıended, if in previous mechanistic studies it was shown that intermittent exposure is the most safe and effective.
10. Estimation of water and food consumption, as well as body weight and temperature should be performed at the same time of the day. During analysis of estrous function, it is necessary to continue to take vaginal smears until the end of the last estrous cycle.
11. For final analysis, it is significant to compare the effects of different doses of drug tested, to extrapolate experimental data to humans.

Applications

DNA mutations have been implicated as causal factors in both cancer and aging and are also responsible for heritable diseases. DNA mutations are irreversible and cannot be repaired in any way other than through the elimination of the cell. Especially large chromosomal mutations are toxic and can lead to genome dysfunction, thereby critically altering normal patterns of gene expression, e.g., through gene-dose and position effects. Somatic mutations in higher organisms have mainly been analyzed using cytogenetic methods or selectable target genes, such as HPRT, in peripheral blood lymphocytes or other actively proliferating cells. Using these approaches, it has been convincingly demonstrated that somatic mutation loads in white blood cells are a major risk factor for cancer and increase significantly with age. Interestingly, in lymphocytes from patients with Werner's syndrome, a segmental progeroid syndrome, the frequency of genome rearrangements was found to be significantly elevated. Meanwhile, nonproliferative tissues, such as liver, heart, and brain, remained inaccessible for detecting random mutational events.

To address this problem, we have developed transgenic mouse models harboring reporter genes, which can be recovered from their integrated state, transferred to *Escherichia coli* and then analyzed for mutations. One of these models, based on plasmids containing the *lac*Z reporter gene allows the detection of a wide range of somatic mutations, including large genome rearrangements. These animals have been studied over their lifetime to test the hypothesis that mutations accumulate with age, as predicted in the original somatic mutation theories of aging.

The results indicate that mutations accumulate in different organs and tissues, albeit at greatly different rates; also the types of mutations that were found to accumulate differed greatly from organ to organ. Interestingly, a significant portion of mutations found to accumulate with age in heart and liver, i.e., up to about 50%, were large genome rearrangements. Such mutations can be detected in the lacZ reporter system as deletions, inversions or translocations when they have one breakpoint in a *lac*Z gene of the plasmid cluster and one breakpoint elsewhere in the mouse genome.

A lacZ mutational reporter system very similar to that in the mouse was recently generated for *Drosophila melanogaster*. The results obtained with this new model are very similar to the results obtained with the mouse, albeit mutant frequencies were somewhat higher with the spectrum dominated by genome rearrangements.

Here, we describe procedures for the quantification and physical characterization of all types of mutations at lacZ reporter loci in both mice and flies. The focus will be on genome rearrangements in view of their potential importance as the most toxic lesions in aging tissues.

Materials

Transgenic animals

1. A lacZ-plasmid C57BL/6J-derived transgenic mouse model used for in vivo mutation detection and characterization has been described. The transgene locus comprises a cluster of head-to-tail, tandemly integrated pUR288 plasmids, manipulated to contain a complete *E. coli lacZ* gene with a flanking *Hin*d III site. Several transgenic lines were derived by microinjection. Line 60 harbors a total of about 20 copies of the plasmid integrated in a head-to-tail fashion at two chromosomal sites: chromosome 3 and chromosome 4. Line 30 harbors 21 copies of the pUR288 plasmid located on chromosome 11. Both lines were bred into homozygosity. The two lines are available upon request from the authors.

2. A similar lacZ reporter system for the detection and characterization of all types of somatic mutations has recently been created for *Drosophila melanogaster*. The transgenic fly lines that contain the lacZ reporter were constructed by P-element mediated germline transformation. The lacZ reporter was cloned into the P-element vector pCasper, which contains the P-element inverted repeat sequences for integration and a portion of the *white* gene for positive selection. This lacZ P-element vector and the P-element vector p(Δ2-3), which contains the transposase gene, were co-injected into embryos that were defective for the *white* gene. The resulting flies were crossed to the *white* minus stock and the progeny were screened for coloration in the eye, denoting successful transformation. p(Δ2-3) is dominantly marked with the *Stubble* gene and negative selection for this marker insures that transposase activity was removed. Individual transgenic flies were again crossed to the white minus stock, to produce a stable stock of that particular lacZ P-element insertion. Several lines of trans- genic flies, each harboring only one copy of the plasmid inserted randomly at different chromosomal locations, were created. All lacZ-fly lines generated are available upon request from the authors.

E. coli strain

The *E. coli* host strain used for plasmid rescue is *E. coli C, ΔlacZ, galE⁻*. It is kanamycin-resistant and sensitive to galactose. The strain is available upon request from the authors.

Reagents

1. 5 X binding buffer: 50 m*M* Tris-HC1 (made from a 1 *M* stock solution, pH 7.5), 5 m*M* EDTA, 50 m*M* $MgCl_2$, 25% glycerol (v/v; 31.5 g/100 mL). Adjust pH to 6.8 with HCl and filter-sterilize the solution.
2. 1X binding buffer: 20 mL of 5 X binding buffer in 80 mL of ultrapure water.
3. IPTG stock solution: 25 mg/mL IPTG (isopropyl β-D-thiogalacto-pyranoside) in ultrapure water. Filter-sterilize and store at –20°C.
4. IPTG-elution buffer: 10 m*M* Tris-HCl (made from a 1 *M* stock solution, pH 7.5), 1 m*M* EDTA, 125 m*M* NaC1. Filter-sterilize and store at room temperature.
5. ATP solution: 10 m*M* ATP (adenosine 5'-triphosphate) in ultra-pure water. Filter-sterilize and store at –80°C in 50 μL aliquots.
6. *Hin*d III restriction enzyme: 10 U/μL (Roche).

7. Restriction endonuclease NEBuffer #2: New England Biolabs.
8. Magnetic beads: Dynabeads M-450 sheep anti-mouse immunoglobulin (Ig)G (4×10^8 beads/ml).
9. Anti-β-galactosidase monoclonal antibody: 2 mg/mL.
10. LacI/LacZ fusion protein.
11. Phosphate-buffered saline without $MgC1_2$: GIBCO BRL.
12. T4 DNA ligase: T4 DNA ligase (Invitrogen; 1 U/μL) is diluted 10X in 1X T4 DNA ligase buffer.
13. Glycogen: 20 μg/μL glycogen (Roche).
14. Sodium acetate: 3 *M* sodium acetate, pH 4.9, in ultrapure water (autoclaved).
15. Luria-Bertani (LB) medium I: LB Broth Base (20 g/L Lennox LB broth base; BD Difco). Autoclave for 20 min (liquid cycle).
16. LB medium II: LB medium I containing $MgCl_2$ in a final concentration of 5 m*M*, added from the 1 *M* stock solution just before use.
17. Magnesium chloride: 1 *M* $MgCl_2$ in ultrapure water (autoclaved).
18. LB topagar: 6.125 g/L LB broth base (BD Difco) and 6.125 g/L antibiotic medium 2 (BD Difco). Add 2,3,5-triphenyl-2*H*-tetrazolium chloride (used as powder to a final concentration of 75 μg/mL; MP Biomedicals) and microwave media until dissolved and then cook for a further 30 min at low power. Allow it to cool down to 44°C before adding ampicillin (final concentration: 150 μg/mL), kanamycin (final concentration: 25 μg/mL), X-gal (final concentration: 37.5 μg/mL) or p-gal.
19. Kanamycin: use as a 50 mg/mL solution.
20. Ampicillin: make a stock solution of 50 mg/mL in ultrapure water.
21. X-gal: 5-bromo-4-chloro-3-indolyl-β-D-galactoside. Use as a 50 mg/mL solution dissolved in *N,N*-dimethylformamide.
22. P-gal: 0.3 % phenyl-β-D-galactoside; use as powder.
23. Lysis buffer: 10 m*M* Tris-HCl, pH 8.0; 10 m*M* EDTA; 150 m*M* NaCl (autoclaved).
24. Potassium acetate: 8 *M* potassium acetate in ultrapure water (autoclaved).
25. X-gal plates: 20 g/L LB Broth Base, 15 g/L agar. Autoclave (121°C, 20 min), cool to 50°C and add 25 μg/mL kanamycin, 150 μg/mL ampicillin, 75 μg/mL X-gal, which are the final concentrations. Pour 35 mL agar per 15 cm-ϕ plate and dry plates overnight at

37°C in the dark. Wrap and store plates at 4°C in the dark for up to 1 mo.

26. TE buffer: 10 m*M* Tris, pH 8.0, 1 m*M* EDTA, pH 8.0 (autoclaved).
27. 10X M9 salts: 58 g/L anhydrous Na_2HPO_4, 30 g KH_2PO_4, 5 g/L NaCl, 10 g/L NH_4Cl.
28. M9CA Medium (1 L): 100 mL 10X M9 salts and 5 g casamino acids in 870 mL redistilled water. Autoclave and allow to cool to room temperature. Add 25 mL 20% (w/v) glucose, 2 mL thiamine (0.5 mg/mL; filter-sterilized), 1 mL 1 *M* $MgSO_4$ and 1 mL 0.1 *M* $CaCl_2$.
29. TY Medium (1 L): 20 g/L bacto-tryptone (Difco), 5 g/L yeast extract (Difco), 5.9 g/L $MgSO_4.7H_2O$, 0.58 g/L NaCl.
30. Lysis solution (7.5 mL): 0.573 g spermidine. HCl, 3 mL 5 *M*NaCl, 0.75 g sucrose.
31. Tris-sucrose buffer (TS buffer): 50 m*M* Tris-HCl, pH 7.5 and 10% (v/v) glucose.
32. Tris-sucrose solution containing protease inhibitors (75 mL): 75 mL TS buffer supplemented with 16 μL aprotinin (5 mg/ml in TS buffer), 16 μL leupeptin (5 mg/mL in TS buffer), 5.2 mg phenylmethylsulfonylfluoride (PMSF) and 80 μL pepstatin (1 mg/mL in TS buffer).
33. 10X DIANA buffer (100 mL): 25 mL 1 *M* Tris-HCl, pH 8.0, 30 mL 5 *M* NaCl, 2 mL 0.5 *M* EDTA, 5 mL 1 *M* $MgCl_2$. Autoclave, allow to cool to room temperature. Add: 0.5 mL Tween20 and store at 4°C.
34. 1X DIANA buffer with glycerol (2 mL): 200 μL 10X DIANA buffer, 1 mL 50% glycerol, 2 μL β-mercaptoethanol, 800 μL distilled water. Store at -20°C.
35. 10% glycerol: add 12.6 g (10X 1.26 g/100 mL) of glycerol in ultra pure water and autoclave.
36. 50% glycerol: add 63 g (50X 1.26 g/100 mL) of glycerol in ultra pure water and autoclave.

Methods

Preparation of electrocompetent cells

1. Add 50 μL of the *ΔlacZ/galE⁻ E. coli* C strain glycerol stock and 5 μL kanamycin to 10 mL LB medium I in a 50 mL Falcon tube. Grow the cells overnight in an Innova 4000 incubator shaker at 37°C at 250 rpm.

2. Add 1.5 mL of the overnight culture to each of four Erlenmeyer flasks containing 500 mL preautoclaved LB medium I. Continue incubation in the shaker, but now at 31.5°C, to an OD_{600} of 0.45.
3. Cool the cells immediately by placing the Erlenmeyer flasks on ice for 30 min on a platform shaker and keep at 0°C from now on as much as possible.
4. Divide the cell cultures over six 500 mL centrifuge bottles (333 mL/bottle).
5. Centrifuge for 15 min at 4000*g* in a Beckman R3C3 centrifuge at 4°C. Gently resuspend each pellet in 167 mL ice-cold distilled water (1 L in total) by shaking gently on a shaker. Do not use force or pipet up and down. Repeat this step one more time.
6. Gently resuspend each pellet in 30 mL 10 % ice-cold glycerol and divide the six suspensions over four 50 mL Falcon tubes. Centrifuge for 15 min at 4000*g*.
7. Gently resuspend each pellet in 2 mL 10 % ice-cold glycerol by shaking gently and combine the contents of the four tubes in a clean Falcon tube. Mix gently by turning the tube.
8. Take 10 μL of the suspension in 3 mL of LB medium I (300-fold dilution) and measure OD_{600}. Adjust density of the concentrated suspension with cold 10 % glycerol such that a 300-fold dilution has an OD_{600} of 0.21.
9. Distribute the competent cells into aliquots of 270 μL in Eppendorf vials and freeze directly in a dry ice/ethanol bath. The electrocompetent cells can be kept at –80°C for several months.

Preparation of lacI–lacZ fusion protein for pUR288 rescue

1. Streak (to isolate individual colonies) the lacI/Z fusion protein producing strain from a glycerol stock on a M9CA medium selective plate (to screen for proline auxotrophy). Grow overnight at 30°C.
2. Select approx 10 colonies and grow these overnight in M9CA medium (1.5 mL). Take samples after a few hours and streak these on TY plates (nonselective plates), containing X-gal (75 μg/mL, final concentration). Continue to grow the colonies and also incubate the plate overnight at 30°C.
3. Select culture that gives highest blue-staining on TY X-gal plate.
4. Inoculate 50 mL M9CA medium with 0.5 mL preculture, and grow overnight at 30°C.
5. Inoculate 8 L of TY medium (non-selective) with 32 mL of the previous overnight culture.

6. Harvest the culture at OD_{595} = approx 1–2 (late log phase). Depending on the exact culture conditions, this value is reached after about 6 h of culturing at 30°C while shaking.
7. Centrifuge the culture at approx 4000*g* for 15 min, and pour off the supernatant. Depending on the number and volume of your centrifuge bottles the centrifuge step can be repeated to accumulate pellet on pellet until the entire 8-L culture has been pelleted.
8. Weigh the pellet(s). The expected yield is 30–40 g.
9. Resuspend cells in 1 mL TS buffer/g pellet weight.
10. Freeze cells by dripping the resuspended cells directly into liquid nitrogen, and store packed frozen droplets (resembling popcorn) at -20°C to keep from desiccating.
11. Preheat a mixture of 75 mL TS buffer containing protease inhibitors and 7.5 mL lysis solution on a heat plate using overhead stirring (do not use a stir bar with magnetic stir plate) to 37°C.
12. Add the frozen droplets (popcorn) swiftly to the preheated (37°C) solution while stirring overhead.
13. Monitor the pH, if necessary adjust with (diluted) HCl to pH 7.5.
14. Add 30 mg of lysozyme (dissolved in minimal volume of 50 m*M* Tris pH 7.5/10% sucrose).
15. Put the suspension on ice for 1 h.
16. Transfer suspension to 37°C water bath for 4 min. And place back on ice.
17. Divide the suspension over centrifuge tubes and centrifuge for 1 h at 23,000*g* at 4°C.
18. Retain the supernatant and determine the volume (~110–140 mL). Place on ice. While on ice, gradually add (during approx 30 min, while gently stirring; use stir bar with magnetic stir plate) ammonium sulphate until 40% saturation (24.3 g per 100 mL volume).
19. Leave on ice (while stirring) for 1–2 h.
20. Divide over 30 mL centrifuge tubes and centrifuge for 15 min, 18,000*g* at 4°C.
21. Store pellets at -80°C.
22. Dissolve one frozen pellet in 1.5 mL DIANA buffer.
23. Store in 100 μL aliquots at -20°C.

Preparation of lacI-lacZ magnetic beads

1. Briefly vortex or shake the Dynal magnetic beads coated with sheep anti-mouse IgG (4 × 10^8 magnetic beads/mL) to ensure that

all beads are resuspended. Pipet 1 mL of the beads into an Eppendorf tube and then place the tube on a magnetic stand (Dynal).

2. Remove the clear supernatant and resuspend the beads in 950 μL phosphate- buffered saline (PBS) plus 50 μL anti-β-galactosidase monoclonal antibody. Vortex the mixture and incubate for 1 h at 37°C while rotating.
3. After incubation, pellet the beads on magnetic stand and remove the clear supernatant.
4. Wash the beads three times with 1 mL PBS and resuspend them in 995 μL PBS plus 5 μL of lacI–lacZ fusion protein. The mixture is incubated for 2 h at 37°C while rotating.
5. Subsequently, the beads are washed three times with 1 mL PBS and resuspended in 1 mL PBS. The lacI–lacZ beads can be stored at 4°C for at least 6 mo.

Mouse tissue collection

1. Sacrifice mice according to local law and institutional policies. In our laborartory this is done by CO_2 inhalation and cervical dislocation.
2. Remove the tissues to be analyzed and rinse them in PBS.
3. Immediately place the tissues in 1.5 mL Eppendorf vials and freeze on dry ice or liquid nitrogen. Maintain samples at –80°C until used.

Fly collection

1. Flies are maintained in plastic bottles with a foam plug at the opening for access to the flies. The flasks are stored at room temperature.
2. To collect the flies, invert the flask and loosen plug slightly to allow entry of CO_2 nozzle.
3. Gently let CO_2 enter the flask to anesthetize the flies.
4. The anesthetized flies are then placed on a plate with continuous flow of CO_2.
5. Using a dissecting microscope and a small watercolor brush, separate the males and females. Male and female flies can be differentiated by examination of the genital organs and coloration. The male genitalia are surrounded by heavy dark bristles, which do not occur on the female. In addition, male flies are completely black in the last one- fifth of the abdomen, whereas females continue the striped banding pattern to the end.

6. Flies are placed in groups of 50 flies per 2 mL Eppendorf tube separated by sex. Tubes are immediately placed on dry ice.

Extraction of mouse genomic DNA

1. Homogenize frozen tissues in 9 mL of lysis buffer in a 50 mL centrifuge tube with a Brinkmann homogenizer. Add 1 mL of 10% sodium dodecyl sulfate (SDS) (1% final concentration), 60 μL of 20 mg/mL RNAse A (120 μg/mL final concentration) and 200 μL of 25 mg/mL proteinase K (0.5 mg/mL final concentration). Digest the tissues overnight at 50°C in a hybridization oven while rotating.
2. The following morning, add 1 volume phenol:chloroform:isoamyl alcohol (25:24:1, w/v) and gently shake the samples on a rocker for 20 min. Centrifuge the samples at 4000*g* for 20 min. Transfer the aqueous fraction to a new tube. Avoid transferring anything from the aqueous/organic interface. Repeat this step two or three times until a clean aqueous phase is obtained.
3. Add one-fifth volume of 8 *M* KAc (approx 2 mL) and mix the samples gently. Add 1 volume of chloroform and gently shake the samples on a rocker for 20 min. Centrifuge the samples at 4000*g* for 20 min.
4. Transfer the aqueous fraction to a clean tube. Avoid transferring anything from the aqueous/organic interphase. Precipitate the DNA by adding 2 volumes of absolute ethanol and gentle mixing. Spool out the precipitate with a sealed glass Pasteur pipet, swish in a tube containing 500 μL 70 % ethanol, and place the DNA into a second tube containing 250 μL 70% ethanol. Centrifuge the tube at 14;000*g* for 5 min and remove all traces of ethanol. Allow to air-dry for a few min. Dissolve DNA in 100–1000 μL TE buffer.

Extraction of fly genomic DNA

1. Homogenize 50 frozen flies in 600 μL of freshly prepared lysis buffer with a disposable pestle in a 2 mL Eppendorf tube.
2. Add 12 μL of Proteinase K (25 mg/mL), 60 μL 10 % SDS and 10 μL RNAse A (20 μg/mL). Mix gently by rotating for 30 min at 65°C.
3. Cool down to room temperature and add 600 μL of phenol: chloroform:isoamyl alcohol (25:24:1, w/v). Rotate the tube for 10 min at room temperature.
4. Centrifuge the samples for 10 min at 10,000*g* at room temperature.
5. Collect the supernatant and transfer to a clean tube containing one-fifth volume of 8 *M* KAc (120 μL) and mix the samples gently

by inverting. Add 1 volume of chloroform and gently shake the samples on a rocker for 10 min at room temperature.

6. Repeat step 4.
7. Transfer the clear supernatant (avoid collecting the interphase) to a 1.5 mL Eppendorf tube and add 2 volumes of absolute ethanol. Gently mix the samples to precipitate DNA. DNA should be easily visible.
8. Centrifuge the precipitated DNA for 10 min at 14,000*g* at room temperature.
9. Wash pellet with 200 μL of 70 % ethanol at 14,000*g* for 5 min.
10. Remove all ethanol and allow samples to dry at room temperature.
11. Resuspend the DNA in 60 μL of TE buffer.
12. Estimate the DNA yield and quality by running a small amount (1 μL) on a 1.5% agarose gel.

Magnetic bead rescue of lacZ plasmid from mouse or fly genomic DNA

1. Take 10–50 μg genomic DNA and make up to 56 μL with ultrapure water.
2. Add 15 μL 5X binding buffer and gently mix up and down with pipet. For the fly, the pre-made mixture contains 15 μL 5X binding buffer with 56 μL of DNA extracted from 50 flies. This mixture can be stored at 4°C.
3. Add 4 μL *Hin*d III (40 U) to the mixture of mouse or fly samples and gently pipet up and down.
4. Pellet 60 μL of lacI–lacZ magnetic beads on the magnetic stand, and discard the supernatant. Resuspend the beads in the DNA/binding buffer/*Hin*d III mix by lightly vortexing. Incubate for 1 h at 37°C while rotating.
5. Prepare two culture tubes for each sample, one containing 2 mL LB medium II and the other containing 1 mL LB medium II. Prepare 1 mL LB medium II per sample for the electroporation and keep on ice.
6. For each sample prepare a mix consisting of 75 μL IPTG-elution buffer, 5 μL IPTG stock solution, 20 μL NEBuffer #2 and 100 μL ultrapure water.
7. After incubation, pellet beads on magnetic stand and discard clear supernatant. Wash the beads three times with 250 μL 1X binding buffer (vortex gently) and add 200 μL of the mixture prepared in

step 6. Vortex gently and incubate for 30 min at 37°C while rotating.

8. Transfer the samples to 65°C for 20 min. Remove the samples and allow them to cool to room temperature before adding 2 μL ATP solution (final concentration: 0.1 m*M*) and 1 μL 0.1X T4 DNA ligase (total amount: 0.1 U). Discard unused portion of ATP.
9. Spin briefly to remove condensation from lid and resuspend the beads by gentle vortexing. Incubate for 1 h at room temperature.
10. Resuspend the beads by vortexing and then place them on the magnetic stand. Transfer the supernatant (200 μL) to a clean tube containing 1.5 μL glycogen (30 μg) and 0.1 volume sodium acetate (22 μL). Vortex the mix and add 2.5 volumes 95% ethanol (560 μL). Mix by inverting.
11. Incubate at –80°C for 20 min and centrifuge for 25 min at 20,000*g*. Remove the ethanol, wash once with 250 μL 70% ethanol (vortex) and centrifuge 5 min at 20,000*g*. Remove all ethanol (use a pipet with a fine tip to remove the last traces) and allow the DNA pellet to dry for 10–15 min. Resuspend the DNA in 5 μL ultrapure water. The electrocompetent cells can now be added.

Electroporation, plating, and mutant counting

1. Thaw electrocompetent cells on ice. Once thawed, directly add 60 μL cell suspension to the tubes containing the 5 μL water/DNA solution. Place on ice.
2. Transfer the cells with the DNA to a prechilled electroporation cuvette on ice. Electroporate at 1.8 kV, 25 μF (Gene Pulser, Bio-Rad) and 200 Ω (Pulse Controller, Bio-Rad). Immediately, add 1 mL ice-cold LB medium II. Then, transfer into the preprepared 15 mL culture tube containing 1 mL LB medium II. Incubate for 30 min at 37°C while shaking (225 rpm).
3. Add 2 μL of the transformed cells (1:1000) to 2 mL LB medium II in a culture tube. Combine with 13 mL LB top agar and carefully pour into a 10 cm Petri dish. This is the titer plate.
4. Add 13 mL LB topagar to the rest of the transformed cells. In addition to ampicillin, kanamycin, and tetrazolium chloride, add p-gal at a final concentration of 0.3 % (add p-gal directly as a powder to the top agar before adding the top agar to the transformed cells). X-gal is not necessary, but can be added also. Plate in a 10 cm Petri dish. This is the selective plate.

5. Both titer and selective plate(s) are grown overnight at 37°C for exactly 15 h, after which colony counting is started without delay. The titer plate indicates the rescue efficiency. A typical yield is about 50,000 colonies (plasmid copies) per microgram genomic DNA. Mutant counting can best be done on a light table.
6. If selective plates are going to be used for mutant characterization, the plates should be wrapped in plastic wrap and stored at 4°C immediately after colonies have been counted.

Data collection

1. Mutant frequencies are determined as the ratio of the number of colonies on the p-gal selective plates (visible as sharp dark-red points) over the number of colonies on the X-gal (nonselective) titer plates (dark-red points with a much larger blue halo) times the dilution factor (usually 1000) of the transformed cells.
2. Each mutant frequency determination point is based on at least three replicates from the same sample, with a minimum of 100,000 colonies for each rescue bringing the total to 300,000 recovered plasmids per sample.
3. The mutant frequency data is imported into an Excel spread sheet to calculate the average and the standard deviation.

Statistical analysis

Statistical analysis depends on the application. The following are examples.

1. *Student's t-test* is performed to test for the significance of a difference between two normally distributed averages from independent samples. A p value of <0.05 is considered significant. Busuttil et al. compared the mean mutation frequencies in proliferating versus quiescent cells using this statistical comparison.
2. *Kruskal-Wallis* test is recommended for statistical analysis of, for example, the effect of donor age when there are 3 or more age groups. If this test shows a significant age effect, several Wilcoxon Rank Sum test should be performed to compare two groups at a time. These nonparametric, analogues of the one-way analysis of variance (ANOVA) and t-tests, require no assumption about the underlying distribution of the data.

Determination of background mutations

While the majority of the recovered lacZ-mutants appear to have originated in the mouse, a subset of mutants are likely to represent artifacts and occur with a frequency of about 1.3×10^{-5}, irrespective of the total mutant frequency. The largest contribution to such artifacts, with a frequency of about 1.1×10^{-5}, originates from "star" activity of the *Hin*d III restriction enzyme during the digestion of genomic DNA in the rescue procedure. Occurring at very low frequency, such star activity results in false deletion mutations. Because all star sites are known, such false-positives are recognized upon restriction analysis of the mutants.

The remainder of background mutants is likely due to spontaneous mutations in the *galK* or *galT* genes, giving rise to galactose-insensitive *E. coli* host cells. This occurs during the preparation of competent cells and the frequency of galactose-insensitive cells can therefore vary from batch to batch. The contribution of both *Hin*d III star mutants and galactose-insensitive cells to the mutant frequency of a sample can be determined as part of the general characterization procedure. Nevertheless, it is important to occasionally assess the background mutant frequency of the lacZ-plasmid system using a "*mock rescue*" of plasmids that have never been part of the mouse. For that purpose we use the following procedure.

1. Isolate *E. coli* C cells harboring the wild-type pUR288 plasmid obtained in the form of a dark blue staining colony on a titer plate from a regular mutant frequency determination.
2. Grow the cells in 3 mL LB medium I containing ampicillin (final concentration: 75 μg/mL) and kanamycin (final concentration: 25 μg/mL) for 8 h at 37°C, 225 rpm.
3. Harvest the cells by centrifugation for 10 min at 1000*g*.
4. Aspirate supernatant and resuspend the pellet in 600 μL of lysis buffer.
5. Dilute the DNA 1:10 and 1:100 in water.
6. Take 1 μL of the diluted DNA and mix with 10 to 40 μg of purified genomic DNA from a non-transgenic mouse or flies.
7. Follow the lacZ rescue procedure as described above to determine the mutant frequency, which should be in the order of 1×10^{-5}.

Mutant characterization

Mutants can be conveniently subdivided into no-change mutants (changes of less than 50 base pairs in size) and size-change mutants (intragenic size changes or genomic rearrangements).

1. Mutant colonies are picked from the p-gal plates and transferred to individual wells of 96-well, round-bottom plates containing 150

μL of LB medium I, 25 μg/mL kanamycin, 150 μg/mL ampicillin per well. Seal the plate and grow cultures overnight at 37°C shaking at 200 rpm.

2. The following morning, examine overnight cultures for growth. Make note of any cultures that did not grow.
3. For each mutant, add 25 μL of a mix containing 11.5 μL ultrapure water, 0.5 μL 12.5 *μM* pUR4923-F primer, 0.5 μL 12.5 *μM* pUR 3829-R primer and 12.5 μL Qiagen HotStar Taq Master mix to a well of a 96-well PCR plate.
4. Transfer a sample of each mutant culture to a well of the 96-well plate containing the PCR mix, using a Boekel replicator. The amplification is carried out at 95°C for 10 min, followed by 35 cycles of 95°C for 20 s, 68°C for 8 min and a final extension of one cycle of 68°C for 10 min.
5. Transfer a sample of each culture with a Boekel replicater (make sure cultures are resuspended) to an X-gal plate for galactose-sensitivity screening. Mark the right orientation on the bottom plate and incubate upside-down at 37°C.
6. Add 50 μL 50 % glycerol to each culture and mix. Seal the plate and store glycerol stocks at –80°C.
7. Examine X-gal plate after 1–2 h for blue staining cultures. These are galactose- insensitive host cells containing wild-type plasmids (false mutants). Make note of blue stained cultures.
8. Continue to incubate X-gal plates overnight at 37°C to screen for mutants. Colorless colonies represent rearrangements or internal lacZ mutations that completely inactivate β-galactosidase activity. Light blue colonies are invariably point mutations, which often only partially inactivate the *lacZ* gene, resulting in some remaining β-galactosidase activity.
9. Digest PCR products with 5 U *Ava* I in NEB#4 (15 μl total volume) in a fresh 96-well PCR plate containing: 5 μL PCR product, 8 μL of water, 1.5 μL 10X NEB#4, 0.5 μL *Ava* I (10 U/μL) and incubate at 37°C for 1 h. Seal PCR samples with HotplateSealer and store at –20°C.
10. Add 2 μL of 10X loading buffer and run digested samples on 1% agarose, 1X TBE gel alongside with 1 kb DNA ladder (GIBCO, 1 μg per lane). Mutants can be conveniently subdivided into no-change mutants (point mutants and other changes of less than 50 base pairs in size) and size-change mutants (intragenic size changes or genomic rearrangements). Classify percentage of no-change and

size-change mutants after having corrected for galactose-insensitive revertants. No-change mutants are classified based on the wild-type restriction pattern of *Ava* I (1992 + 1443 + 818 bp) on 1 % agarose gel. Any deviation from this pattern indicates a size-change mutant.

11. To check for possible cloning artifacts, each mouse sequence mutant is routinely redigested with *Hin*d III. Occasionally, random *Hin*d III fragments of the genome are cloned into the *Hin*d III site of the plasmids. The cloned fragment does not alter the phenotype of the bacterial host, but could make point mutants to appear as size-change mutants in the *Ava* I characterization described above. A *Hin*d III-digested PCR-amplified *lacZ* gene showing more than two fragments typically indicates a cloning artifact.

Quick confirmation of genome rearrangement mutations by PCR

This PCR test can be used in order to classify size change mutants into those which are internal (i.e., occur entirely within the *lac*Z gene) or those which involve the mouse genome. It does not give the exact breakpoints of the mutation (sequencing is required for this). This PCR amplification was designed to use a forward primer for the last 18 bp of lacZ prior to the *Hin*d III site (pUR3 285-F) in combination with a reverse primer after the *Hin*d III site (pUR3421-R). This primer combination results in a 137-bp fragment, only when the 3' end of the *lacZ* gene is still present. Hence, the absence of a product indicates a 3'-deletion resulting from a large rearrangement. As a positive control for the PCR, a primer set specific for the ampicillin resistance gene (pUR4071-F and pUR4181-R) is used to co-amplify a 111-bp fragment in the same reaction. Mutant plasmid templates resulting in an amplified 11 1-bp fragment and an absent 137 bp fragment are assumed to have a truncated *lac*Z gene as the result of a rearrangement mutation, having one breakpoint in the mouse/fly genome. Note that star-mutants discussed above will appear as genome rearrangements in this test. This protocol is always used on mini-prepped mutants and not on PCR-amplified plasmid sequences. In our experience, mini-prepped lacZ-plasmids provide a better template for subsequent sequencing. If PCR is used, new primers must be designed.

1. Miniprep mutants following the manufacturer's protocol for low copy plasmids.
2. Prepare a PCR mix containing 12.5 μL of Qiagen Hotstart mix, and 1 μL of each of the primers at a starting concentration of 12.5 μM and 8 μL water.

3. Add 0.5 μL of template DNA. The amplification is carried out at 95°C for 12 min, followed by 35 cycles of 95°C for 40 s, 50°C for 40 s, 72°C for 40 s and a final extension of one cycle of 72°C for 10 min.
4. Add 2 μL of 10X loading buffer to each PCR product and load entire sample on a 2% agarose, 1X TBE gel.

Mutant sequencing

To identify the nature of a lacZ mutation or the breakpoint of a rearrangement, sequencing is necessary. As mentioned above, this can best be done on mini-prepped plasmids.

1. Select the mutants of interest and inoculate 6 mL of LB medium I containing 25 μg/mL kanamycin and 150 μg/mL ampicillin. Because pUR-228 is a low copy plasmid, three miniprep culture tubes per sample are necessary to obtain the required amounts of plasmid DNA for sequencing the entire *lac*Z gene.
2. Grow samples overnight at 37°C at 225 rpm.
3. The following morning, spin the samples at 272*g* at room temperature.
4. Aspirate the LB medium I from the culture tubes being careful not to disturb the cell pellet.
5. Follow the manufacturer's protocol for mini-preparation of low-copy plasmids (QIAprep Spin Miniprep kit/Qiagen).
6. Elute the samples in 50 μL of ultrapure water and combine samples from the three tubes.
7. Purify samples using Montage PCR Centrifugal Filter Devices (Millipore) and resuspend the concentrated plasmid DNA in 45 μL of ultrapure water.

The primers designed to cover the entire *lac*Z gene from base 1 to 3096 with reference to the pUR288 plasmid sequence available in Genbank. In order to determine the nature and position of point mutations, it is necessary to sequence the entire *lac*Z gene, in which case all primers need to be used. Primers to sequence size-change mutants are selected on a case by case basis according to the *Ava* I restriction pattern of the mutant (not all primers are required for all mutants). Mutations are identified by comparing the sequenced mutant to the pUR288 consensus sequence. Using the mouse or fly sequence database the 3' end breakpoint in the mouse or fly genome can be determined on the basis of the small mouse or fly fragment captured as part of the lacZ-mutant plasmid.

Both line 60 and line 30 lacZ transgenic mice harbor known polymorphic sites, which should not be confused with mutations when comparing mutant with wild-type lacZ sequences. The polymorphic sites are always accompanied by a unique mutation. The transgenic *Drosophila* lines developed in our laboratory contain no known polymorphic sites.

For the line 60 mice, fluorescent *in situ* hybridization (FISH) and genetic mapping were used to determine that the plasmid concatamers are positioned 59.5 and 87.2 Mb from the centromere on chromosome 3 and 4, respectively. For line 30, FISH was used to determine that the *lac*Z gene is localized to chromosome 11, A1–A2 region. However, the exact physical location of the lacZ-plasmid cluster is not currently available for th'- mouse line, because its integration site has not been cloned. For the transgenic fly, a total of 12 different lines were generated but only for line 11B, with its integration site on chromosome 3R, the integration site has been mapped and the 5'-sequence immediately adjacent to the transgene construct characterized. In this case, the pCasper vector was found to be located 913 kb from the centromere.

Notes

1. It is important that all reagents are absolutely clean and contain no traces of β-lactamase-carrying plasmids (the aim of the assay is to select for plasmids with dysfunctional *lac*Z gene conferring ampicillin resistance).
2. Preparing lacI–lacZ fusion protein:
   (a) Maintenance of plasmids bearing the lacI–lacZ fusion protein requires minimal medium. The plasmid is maintained by complementation of a proline auxotrophy in the strain.
   (b) When adding the ammonium sulfate too fast, one risks high concentrations localized near the slow dissolving ammonium sulphate which could result in precipitation of fusion protein.
3. Extraction of mouse genomic DNA:
   (a) Some tissues (small intestine, spleen, and kidney) have a tendency to give a very viscous aqueous phase. In this case, additional phenol and chloroform extractions are required.
   (b) Extracting DNA from a nontransgenic tissue in parallel with regular samples is always recommended as a control against contamination.
   (c) It is not always necessary to extract DNA from the entire organ. For example, one-third of liver, one kidney, one small

piece of the small intestine, half of a spleen will each yields sufficient DNA to perform the assay multiple times.

4. Magnetic bead rescue of lacZ-plasmid from mouse or fly genomic DNA:
   (a) For viscous DNA, increase the reaction volume for the same amount of DNA with 1X binding buffer.
   (b) Individual lacZ reporter plasmids in mouse and in fly samples can also be recovered using *Pst* I restriction enzyme. In contrast to *Hin*d III, *Pst* I digestion of genomic DNA excludes the recovery of genome rearrangements with one breakpoint in the reporter gene and one elsewhere in the genome; only intragenic events, e.g., deletions, insertions, and point mutations, can be recovered. Because of the location of the *Pst* I restriction site in the β-lactamase gene, mutant plasmids resulting from genome rearrangements are excised with incomplete ampicillin resistance genes. This explains why lacZ plamids excised with *Hin*d III show much higher mutation frequencies than after *PstI* digestion.
   (c) Recently, we have changed *Hin*d III suppliers from New England Biolabs to Roche. Although earlier batches of the NEB enzyme have performed well in this mutation assay, the most recent batch we and others obtained from New England Biolabs resulted in additional background colonies, carrying non-pUR288 derived plasmids. Switching suppliers solved this problem. As concentrations between the suppliers differ by 2-fold we adjusted the volume of *Hin*d III accordingly with respect to previous publications.
   (d) Several controls should always be added as extra samples in the rescue procedure: (1) a nontransgenic control serves as a control against contamination encountered during the DNA extraction process; (2) a water control should be included to identify any contamination which arises during the rescue procedure either as a result of experimental error or contaminated solutions; (3) one sample with known mutation frequency values and known efficiency in the X-gal plate should also be included as a reference for the new rescue values found in each assay.
5. Mutant characterization. For the mouse, it is now necessary to correct for so-called *Hin*d III "star" mutations. The original lacZ-plasmid used to generate the mouse models contained several of

these sites, which at very low frequency become subject to digestion, falsely indicating size-change mutations. These false positive mutants usually do not make up more than 20% of total size changes. Of note, the *Drosophila* model does not contain such sites, which were removed from the lacZ construct for generating this model (Garcia et al., submitted).

6. Overnight incubation time prior to counting the colonies. It is important to keep the overnight incubation period of the transformed cells identical to allow comparing results between different days. In particular, the selective plates contain a high concentration of nontransformed cells. In our experience these nontransformed cells are not efficiently killed by ampicillin. After the ampicillin is sufficiently degraded, some of the nontransformed *E. coli* cells start to form so-called satellite colonies around the true mutant colonies. Because this is a gradual process, variations in mutant frequencies are introduced when incubation times differ between experiments. However, in practice colony counting of large experiments may take a significant amount of time and sometimes interfere with other experiments planned that day. Because the problem is most pressing for the selective plates, as indicated previously, we advise counting the selective plates first. Should another experiment require attention first, one can place the plates at 4°C for a few hours to delay counting.

14

Biogerontology

It is widely known that *caloric restriction* (CR) remains the only nongenetic and most robust intervention that reproducibly extends both average and maximal lifespan in a wide variety of species, including mammals. An important distinction to be drawn is that CR, unlike other approaches, increases both average and maximal lifespan. A number of different interventions can act to extend average lifespan in both animals and humans. Consider that average lifespan for men and women has increased remarkably since the early 1 900s in developed countries from about 40 yr then to 80 yr today.

This is mostly attributable to a reduction in early mortality due to improvements in medical care (antibiotics, vaccinations, and so on), nutrition, and improvements in overall health. Despite notable increases in the average human lifespan, the maximal (presumably genetically determined) lifespan has remained fixed around 120 yr. It can be argued that interventions that affect average lifespan do so by altering pathologies associated with aging and age-related disease, whereas interventions such as CR that also increase maximal lifespan fundamentally alter the underlying biology of aging.

As such, CR is not only the most robust intervention for lifespan extension and retardation of aging, but may also be an important tool available to scientists to explore fundamental biological mechanisms of aging.

Experimental Models of Caloric Restriction and Applicability to Humans

The lifespan-extending and other beneficial effects of CR, such as anti-tumor effects and the maintenance of more youthful physiology, have been reported in many hundreds of experiments over the past 70 yr. Nonetheless, the question of relevance to humans remains and will go unanswered until definitive human data are obtained. There are, however, data from a number of sources that suggest that CR may be relevant to human aging. For example, based on his study of Spanish nursing home residents, Vallejo concluded that reduced caloric intake was associated with a significant reduction in morbidity and that mortality also tended to be lower in the group provided the fewest calories. Caloric intake in residents of the Japanese island of Okinawa differs by 20–40% in adults and children, respectively, compared to the national average. Interestingly, Okinawa has a greater proportion of centenarians, a lower overall death rate, and fewer deaths due to vascular disease and cancer. Although these were largely uncontrolled studies, they do suggest a possible link between calorie intake, disease, and perhaps even longevity.

More controlled studies of CR are ongoing using nonhuman primates, mostly Rhesus monkeys. Given the phylogenetic proximity between Rhesus monkeys and humans, these studies will shed some light on the question of CR relevance to man. Data from the primate studies have been reviewed elsewhere and will not be discussed in detail here. Briefly, monkeys on CR exhibit many physiological responses consistent with those observed in rodents such as reductions in body weight and adiposity, fasting glucose and insulin, body temperature, and changes in serum lipids (e.g., cholesterol and triglycerides), among others. Emerging data suggest that CR may also favorably impact age-related diseases and associated morbidity and perhaps even mortality.

Another link between CR and human aging comes from studies of men in the Baltimore Longitudinal Study of Aging. Along with other National Institute on Agin (NIA) colleagues, we investigated whether three of the most robust physiological markers of CR in animals were related to survival in men in this ongoing study. In a non-CR cohort of normal men, reduced fasting insulin and body temperature and maintenance of higher levels of the adrenal steroid, dehydroepiandrosterone-sulfate (DHEAS),were associated with greater survival. Finally, short-term studies in humans show that two of these three markers (reductions in fasting insulin and body temperature) are observed in response to short-term CR. Taken together, these findings bode well for the possibility that CR may indeed extend lifespan and retard many of the pathological processes associated with advancing age.

Table 14.1. Effects of calorie restriction on selected parameters of morphology, physiology, aging and disease in Rhesus monkeys

Category/parameter	*Decrease*	*Increase*	*No change*
Body Composition			
Body weight	X		
Fat and lean mass	X		
Trunk: leg fat ratio	X		
Height	X		
Development			
Time to sexual maturity		X	
Time to skeletal maturity		X	
Metabolism			
Metabolic rate (short-term)	X		
Metabolic rate (long-term)			X
Metabolic rate (long-term:nighttime)	X		
Body temperature	X		
Thriiodothyronine (T3)	X		
Thyroxin (T4)			X
Thyroid Stimulating Hormone (TSH)			X
Leptin	X		
Endocrinology			
Fasting glucose/insulin	X		
IGF-1/Growth Hormone	X		
Insulin sensitivity		X	
Age-Related Maintenance of Melatonin and DHEAs		X	
Testosterone; Estradiol			X
Cardiovascular Parameters			
Systolic blood pressure	X		
Heart rate	X		
Serum triglycerides	X		
Serum HDL2B		X	
LDL interaction with proteoglycans	X		
Lipoprotein(a)	X		
Immunological Parameters			
IL-6	X		
IL-10	X		
Interferon-γ		X	

Oxidative Stress			
Oxidative damage to skeletal muscle	X		
Cell Biology			
Proliferative capacity of fibroblasts		X	
Glycation products	X		
Functional Measures			
Locomotor activity			X
Acoustic responses		X	

To achieve the maximum benefit from CR, presuming its applicability, humans would need to reduce their caloric intake by about 30% or from approx 2500 calories per day to 1750 (in men). The tremendous popularity of diet books, pills, and associated weight-loss products illustrates the challenge that use of such a regimen on a wide scale would present. Thus, to achieve the potential benefits of CR, an alternative approach is needed. In considering possible biological mechanisms of CR, we hypothesized that, by targeting alteration of cellular energy metabolism, it might be possible to "*trick*" the body into shifting to a CR-like survival mode without actually reducing food intake, and in this way "*mimic*" the effects of CR.

Caloric Restriction Mimetics

We first proposed the idea of CR mimetics in 1998 and further expanded on this potential approach in a subsequent article in *Scientific American*. In our initial study, we reported that disruption of cellular glucose metabolism (e.g., *glycolysis*) using the glucose analogue 2–deoxy-D-glucose (2DG) fed in the diet to rats lowered body temperature and fasting insulin levels without significantly reducing food intake over a 6-mo period at the selected dose. The 6-mo duration of this study was insufficient to assess indices of biological aging or longevity, but did validate that it may be possible to "*mimic*" metabolic effects of CR without reducing food intake. A follow-up survival study in rats unfortunately indicated that the window between efficacy and toxicity was too narrow to make this particular compound useful. The concept of CR mimetics has been further validated in other experiments. For example, similarly to CR, 2DG has been shown to be neuroprotective in rodent models of neurotoxicity and ischemia, and disruption of a different enzyme in glycolysis by iodoacetate protects neurons from glutamate toxicity. Thus, it is possible to mimic metabolic and protective effects of CR without reducing normal food intake.

Use of the term CR mimetics is becoming increasingly commonplace in the gerontological literature, with CR mimetics often loosely defined as any intervention, genetic or otherwise, that results in lifespan extension in a fashion similar to CR. In many cases, it is not understood whether food or energy intake were reduced by the intervention being tested, making it impossible to determine whether any effects observed mimicked CR or were in fact due to an actual reduction in energy intake (i.e., CR). In our view, a CR mimetic must have limited, if any, effects on food intake and must specifically target energy metabolism, because it is well established that energy is the nutritional factor responsible for the diverse and beneficial effects of CR. Another critical point is that to mimic CR effectively and to have a potential impact on fundamental processes of aging, candidate mimetics must increase both average and maximal lifespan. These important points are raised not in an attempt to limit the scope of possible CR mimetic approaches, but to ensure that candidate CR mimetics are appropriately focused and evaluated.

Possible Metabolic Targets for CR Mimetics

CR mimetics have been proposed that target a number of pathways related to energy metabolism such as glycolysis inhibitors, antioxidants, sirtuin regulators, and insulin sensitizers. As summarized previously, inhibition of glycolysis remains a promising target despite our disappointing results with 2DG. Given the popularity and general acceptance of the Free Radical Theory of Aging, antioxidants have been the focus of many studies in biogerontology.

Treatment with antioxidants has occasionally been shown to lead to an increase in average lifespan; however, reproducible increases in both average and maximal lifespan in mammals have not been observed.

Sirtuins serve as gene silencers, and emerging evidence supports a possible role in aging and lifespan extension in short-lived organisms. Insulin sensitizers may also act to mimic CR, because an increase in insulin sensitivity is among the most rapid and robust response to this nutritional intervention. In vitro studies have shown that phenformin, a treatment for diabetes, suppresses calcium responses of hippocampal neurons to glutamate and decreases their vulnerability to excitotoxicity. Studies of phenformin treatment in vivo on markers related to CR and lifespan have been inconclusive. However, at least one study has shown that phenformin reduced reactive oxygen species.

Given the redundancy of metabolic pathways and their sometimes differing regulation in different tissues, it seems unlikely that a CR mimetic targeting a single pathway will produce all of the beneficial effects of CR. It may be necessary to target, for instance, both glucose and lipid metabolic pathways to achieve the full benefit of CR without reducing food intake. Thus, it is conceivable that "*cocktails*," containing various combinations of candidate "*segmental*" CR mimetics, might be devised to more completely duplicate the effects of CR.

Beyond the obvious potential benefits of CR mimetics on lifespan and aging, this approach provides additional opportunities to probe aging at the molecular level. The utility of CR as such a probe is limited by the fact that it requires energy restriction at the level of the whole organism from which tissues can then be harvested for study. Given this constraint, certain types of molecular studies are challenging at best or at worst not possible. If further validated, this approach has obvious benefit in that it provides molecular gerontologists with additional tools to probe biological processes of aging at the subcellular level, perhaps ultimately leading to the development of pharmaceuticals that may also mimic CR.

CR remains the most robust and only reproducible intervention for extension of both average and maximal lifespan and, as such, for altering the fundamental biology of aging. In addition, CR consistently slows many physiological and pathological changes that occur with advancing age thereby maintaining health and vitality. Recent experiments with both monkeys and humans suggest that this energy restriction paradigm may indeed impact on human aging. Even if CR is shown to be an effective intervention into human aging processes, it is unlikely that the 30–40% reduction in calories over much of the lifespan necessary to achieve maximal benefit will be practiced effectively on a wide scale. Thus, development of CR mimetics can provide an alternative strategy for slowing human aging.

Early experiments with glycolytic inhibitors and other compounds have validated the potential application of this approach and have helped to broaden the search for true CR mimetics. It seems unlikely that a single candidate CR mimetic will produce all the beneficial effects of actual CR. Therefore, "*cocktails*" containing combinations of these substances appear to be feasible and may ultimately provide the most successful approach. CR mimetics will also likely provide biogerontologists with an important tool for investigating molecular mechanisms of aging.

15

PROTEIN IN AGING

The basic methodology of two-dimensional gel electrophoresis (2DE) is several decades old, but recent technological advances in both 2DE and *mass spectrometry* (MS) have dramatically increased the impact that protein electrophoresis can have in modern biomedical research. In particular, commercially available plastic-backed strips of low percent acrylamide containing immobilized pH gradients in several ranges have made *isoelectric focusing* (IEF) accessible to any careful researcher. The advances in MS and computer technology have allowed researchers to couple the power of MS technology to different types of up-front protein or peptide separations to enable high-throughput analysis and identification of polypeptides, their proteolytic fragments, and their modifications, in biological samples. These *proteomic approaches* are enabling the acquisition of information about protein expression and/or modifications in biological experiments to an unprecedented extent currently.

We utilized 2DE gel proteomics to identify proteins involved in the purported health benefits of the dietary supplement GSE, marketed for anti-oxidant activity; young adult rats were chosen for the initial study to determine whether a dietary supplement taken early in adult life could have ramifications for late life health, and during aging. Much of the 2D gel and proteomic methodologies described here are taken from that study. Because of the purported importance of protein oxidations in aging-related diseases, this chapter describes Western blot procedures to detect and quantitate protein oxidations in biological samples, as well as the MS methods to identify the amino acid residues involved in the oxidations. Finally, 2D blue native electrophoresis methodology is described, which enables separation of protein complexes in a nondenaturing first dimension, and then the resolution of the components of such complexes in a denaturing second dimension electrophoresis. This chapter focuses on *2D electrophoretic* protein separation methods prior to MS analysis. Other overviews discuss other separation technologies such as MUDPIT, stable isotope labeling technologies such as iCAT and metabolic labeling, as well as other MS approaches, including top-down and fourier transform-ion cyclotron resonance MS. The volume of data that MS technologies generates is such that duplication of the experiment is rarely carried out, resulting in matches of peptide masses and even amino acid sequences to proteins that are not necessarily valid; these appear in the literature, however, because careful reviewing of such data itself requires familiarity with such technologies, which is not yet sufficiently widespread among researchers carrying out the biological experiments.

The proteomics technologies available today generate such volumes of data that the mining of the data, and which data to archive, are nontrivial. Quality control issues become important considerations: suggestions as to how quality control issues pertinent to different parts of the methods can be addressed. In the end, proteomics technology is too expensive and time-consuming to expend on experiments in which bias from different sources and technical and day-to-day variability have not been adequately addressed.

Materials

Chemicals Needed for 2DE and their Sources

1. Amberlite IRN-150L.
2. DryStrip Cover Fluid (mineral oil).
3. Immobiline Drystrips, pH 3.0–10.0, 11 cm, 12 pack.
4. Sypro Ruby gel stain (liquid, room temp, protect from light).
5. Acetic acid (reagent grade, room temp).
6. Ampholytes (100% solution, 4°C).
7. Bromophenol Blue (powder, r.t.).
8. CHAPS (powder, -20°C).
9. Colloidal Coomassie Brilliant Blue gel stain (liquid, 4°C).
10. Dithiothreitol (DTT) (powder, –20°C).
11. Glycerol (87% liquid, room temp).
12. Glycine (powder, room temp).
13. Iodoacetamide (powder, –20°C).

14. Protein assay reagent (liquid, room temp).
15. Sodium dodecyl sulfate (SDS) (powder, room temp).
16. Tributylphosphine (TBP) (liquid, ampoules, -20°C).
17. Thiourea (powder, room temp).
18. Tris HCl (powder, room temp).
19. Urea (powder, room temp).

Equipment

1. Polyethylene screw-cap tubes, 15 mL, 50 mL.
2. Glass Dounce homogenizers; 2 mL, 7 mL, 10 mL.
3. Ettan IPGphor II isoelectric focusing (IEF) unit.
4. Immobiline Drystrip Reswelling Tray.
5. Ceramic boats for IPGphor (for 11-cm strips).
6. Optima TLA 100 tabletop ultracentrifuge.
7. TLA 100.2 fixed angle rotor.
8. Thick-walled polycarbonate tubes for TLA 100.2 rotor.

Stock Solutions

10% (w/v) SDS

1. Fill a 100 mL screw-cap bottle with premeasured 100 mL of double-distilled H_2O, and add a clean stirring bar.
2. Mark the bottom of the meniscus with a magic marker, then pour out about one third of the volume into another clean beaker.
3. Weigh out 10 g SDS with care, and dispense gradually directly into the screw-cap bottle containing the H_2O with stirring.
4. This concentration of SDS must be warmed to at least 37°C for complete dissolution.
5. When completely dissolved, adjust the water to 100 mL, and stir again to ensure homogeneity. Store capped at room temperature.

10 M Urea

1. Fill a clean 1 L glass graduated cylinder with MilliQ water about two-thirds full, and start it gently stirring.
2. Weigh 601 g urea in a large weighing boat, and pour it into the cylinder in a fine stream, stirring steadily so that the urea starts dissolving immediately.
3. Add more water, stirring until the urea is dissolved, then adjust with water to the 1-L line, *with the stirring bar stopped*. Then stir it again to make sure it is homogeneous.

4. After the urea is completely dissolved, add about 10 g of the Amberlite beads and stir gently for 1 h. This deionizes and removes charged contaminants from the solution.
5. Finally, sterile-filter the solution using a 0.22 μm filter, making sure that no Amberlite beads go into the final bottle. Keep the urea solution capped at room temperature. Do not expose to temperatures above 30°C.

1.5 M Tris, pH 8.8

1. For 0.5 L, weigh out 90.8 g Tris base.
2. Calibrate the pH meter prior to making this solution; because many solutions are pH'd near physiological pH (7.0–7.5), the meter may need fresh calibration at pH 7.0 and at pH 10.0.
3. Mark a 0.5 L glass beaker with a felt tip marker with a thin line running along the bottom of the meniscus when filled with 0.5 L of H_2O and a stirring bar. *Then pour out about one third of the* H_2O.
4. With the H_2O stirring, add the Tris base gradually to the beaker. Let it dissolve completely and check the pH. It should read somewhere around pH 10.0–11.0.
5. While stirring, with the pH meter on and the electrode in the solution, add drops of concentrated HCl to the Tris solution, until the pH is adjusted to 8.8.
6. Then adjust with H_2O to the appropriate volume mark without stirring, then *stir again to ensure homogeneity*, and sterile filter into a 0.5-L screw-cap bottle. Keep at 4°C.

Table 15.1. Isoelectric focusing lysis buffer

Component	*Desired concentration*	*Stock solution*	*Amount of stock for 50 mL*
Urea	7 *M*	10 *M*	35 mL
Thiourea	2 *M*	powder	7.61 g
CHAPS	4%	powder	2.0 g
Tris-HCl	40 m*M*	1.5 *M*, pH 8.8	1.33 mL

IEF lysis buffer

1. Mix components listed in Table 15.1 in water with stirring, and adjust to 50 mL with water, making sure that everything is dissolved.
2. Sterile-filter through a 0.22 μm filter as before into a sterile 50 mL polyethylene tube.

3. Aliquot into 10 mL portions, label with date, and keep at –20°C.
4. For use, thaw one tube, aliquot it into 1 mL portions, and use these up before thawing out another 10 mL portion.
5. Just before use, add TBP and ampholytes (in the same pH range as the IPG strips), to make it 5 m*M* TBP (25 μL of 200 m*M* TBP stock solution), and 0.5% (v:v) ampholytes (5 μL 100% stock solution per 1 mL lysis buffer [LB]).

IEF rehydration buffer

1. Mix components for IEF Rehydration Buffer.
2. Bring to volume (50 mL) with water, making sure that everything is dissolved.
3. Sterile-filter through a 0.22 μm filter as for the LB into a 50 mL polyethylene tube.
4. Aliquot in 10 mL aliquots in 15 mL polyethylene screw-cap tubes at –20°C.
5. As with the LB, adjust with TBP and ampholytes (in the same range as the IPG strip being used, i.e., 4-7 or 3-10) just before use.

Second-dimension equilibration buffers

1. Mix components as for IEF solutions.
2. Bring to volume (200 mL) with water, making sure that dry components are dissolved.
3. Sterile filter through a 0.22 m filter into 50 mL polyethylene tubes; store at –20°C.
4. For reducing equilibration buffer (EqB) (65 m*M* DTT) (DTT-EqB): add 0.1 g of powdered DTT fresh to 10 mL of EqB.
5. For alkylating EqB (2.5% Iodoacetamide (IAA-EqB): add 0.250 g of IAA and 2 μL of 10% Bromophenol Blue fresh to 10 mL of EqB. IAA is light sensitive; when making and using this buffer, protect it from light.

10X SDS-PAGE running buffer (1.92 M glycine/250 mM Tris, pH 8.3)

Make 10 L at a time in a clear carboy with a stirring bar and spigot and keep at room temperature.

1. Dissolve 450.2 g glycine and 302.8 g Tris base in a final volume of 10 L of double distilled H_2O. The unadjusted pH of this solution is 8.3, which does not need to be adjusted further.
2. However, a drop should be checked with pH paper to make sure there have not been errors.

3. *Note*: this 10X stock solution does not contain SDS. We add the SDS only to the running buffer for the upper reservoir, since it is not required in the bottom buffer.

Methods

Preparing the Sample for IEF

1. Sample choice: this chapter deals with how to analyze experimental biological samples using proteomics methods, and presumes an experiment has been carried out to generate biological samples that hypothetically contain protein differences between control and stimulated/treated, or between wild-type and mutant/transgenic cells or animals. For our previously published study, whole rat brains were dissected out of the skull following sacrifice of the animal, weighed, snap frozen in liquid nitrogen, and stored at -80°C until used. Similar processing would be followed for other tissues/organs. Cell pellets could be processed similarly, following several washes with ice-cold phosphate-buffered saline (PBS) to get rid of cell culture medium.
2. Snap-freeze the tissue or cell pellet by submersion in liquid nitrogen. For tissue, first weigh it on a plastic weigh boat, then wrap it and the boat with foil, pre-labeled with identifying information on tape resistant to -80°C, and then finally submerge in liquid N_2 until it sinks. Once all tissues have been treated similarly, store them in a -80°C freezer.
3. For IEF, take a brain out of the -80°C freezer, place the tissue in a heavy-duty plastic bag and shatter the tissue with a hammer to knock off an appropriate size piece or thaw the brain sufficiently to make a sagittal cut, and return the rest of the tissue to the -80°C freezer. For analysis of events affecting the whole brain, a sagittal cut is optimal. Half of a rat brain provides ample amounts of tissue for 2D gel purposes, so that the other half can be saved for further or alternative analyses. Re-weigh the tissue. For studies addressing aging-related cognitive impairment such as memory deficits, brain regions such as hippocampus or frontal cortex should be studied. These latter dissections are better accomplished with brain freshly dissected out of the animal, vs after the brain has been frozen and stored at -80°C.
4. Homogenize the tissue in a glass: glass homogenizer in LB at a tissue:buffer ratio of 1:5 (wt:vol) at room temperature until the homogenate appears homogeneous (about five passes with the

pestle), then do five more passes. For tougher tissues such as heart or mammary gland, a glass:Teflon motorized homogenizing system is necessary, especially for multiple samples.

5. Allow the homogenate to sit at room temperature for 30 min, then re-pass the pestle through the homogenate, to make sure all possible "macro"-aggregates are disrupted.
6. Clarify by centrifugation at 100,000*g* for 30 min, in the TLA 100 ultracentrifuge. The settings are 55,000 rpm, 30 min, 22°C, with braking at the end of the spin. Make sure this spin is not done in the cold; this will cause the urea to crystallize, causing uneven solubilization of the proteins.
7. Collect the supernatants carefully and pool (if more than one tube), mixing gently - then re-aliquot in small (100 μL) volumes and store at –80°C.
8. Set aside one small aliquot (10 μL) of each pooled sample for a protein concentration determination. Our lab uses the BioRad Bradford assay reagent, but other brands can suffice. Brain homogenized using the tissue:buffer ratio of 1 g:5 mL typically yields a concentration of approximately 10 mg/mL. For a given type of sample, if care is taken to quantitatively homogenize all samples similarly with the same buffer, all samples should yield similar concentrations within an experiment.

First Dimension: Isoelectric Focusing

1. Using the protein assay information, calculate for each sample to be analyzed the volume required for 100 μg protein. For 11-cm IPG strips, our most typical load is 150 μg, although 50 μg suffices for analytical purposes for many samples. The optimal protein load for a sample not previously analyzed should be determined empirically by loading replicate gels of different amounts of the same sample and going through the entire procedure. If the sample is a tissue or whole cell pellet, and it is suspected that the proteins of interest are low abundance, 500 μg may be necessary. A trial run of three strips of one representative sample containing 50, 100, and 200 or 500 μg is recommended if there is sufficient sample. If insufficient sample prevents these loadings, then work with the highest two volumes possible, such as 250 and 500, but note them for future reference.
2. Dilute this sample with additional LB to 110 μL and then dilute 1:1 with RHB to 220 μL, mixing gently by pipetting up and down with a 200 μL pipet.

3. Carefully pipet 200 μL of this into a trough in the rehydration tray, keeping it along the middle of the trough, in a continuous stream.
4. At this point, take the IPG strips out of the –80°C freezer, and lay them one at a time gel-side-down in the appropriate trough, starting at one end, so that the sample volume is underneath the strip and not being pushed out the sides of the strip. Lift and replace each half of the strip in turn to maximize contact with the sample and eliminate bubbles. There should not be a discontinuity of sample anywhere in the strip, indicated by a discontinuity of blue down the strip; if this occurs, the strip may "burn," resulting in a faulty IEF run. If a discontinuity of blue sample is observed in the middle of the strip, use a pipet tip to shift remaining sample in the trough to that region of the strip, and lay the strip back down. Now lift the strip again to ensure that the blue is continuously down the whole strip. If a discontinuity persists, process more sample and pipet it directly underneath the affected length of strip; although the precise protein load for that strip is now different than the others, it is preferable to getting no information from that strip as a result of not running it, or to having it burn.
5. Once one strip has been processed as in step 4, overlay it with mineral oil to prevent urea from crystallizing.
6. Repeat this process (steps 3–5) for the remaining samples until all have been overlaid with oil.
7. Cover the rehydration tray with the cover and let the strips rehydrate overnight (at least 12 h) at 22°C.
8. The next morning, remove the strips from their troughs, rinse lightly with water, and place in the Ettan IPGphor IEF "boats," taking care to orient the positive end of the strip at the positive end of the boat.
9. The wicks for the IEF should be damp with double distilled H_2O, *but not dripping*, and placed just barely inside the electrode wires to ensure complete conductance to the gel on the strip.
10. Set the maximum current per strip at 50 μA and the temperature at 20°C.
11. With the IPGphor as with other IEF units, the parameters for a run can be programmed. The IEF will stop automatically.
12. At the end of the run, remove the strips from the boats, and drain excess mineral oil by horizontally touching the strip briefly to filter paper.

13. Store the strips individually in 15 mL polyethylene conical screw cap tubes, sliding the plastic side of the strip in contact with the inside of the tube to prevent disturbing the gel side. Lay the tubes horizontally in the –80°C freezer with the IPG strips at the bottom, and protect them from being physically disturbed.

Reduction and Alkylation, and Equilibration for SDS-PAGE

1. Take the IPG strips out of the –80°C freezer, and place them in a trough in the equilibration tray; fill with freshly made DTT-EqB; rock gently on an orbital shaker for 15 min, making sure there is no spillage between adjacent troughs.
2. Transfer to fresh DTT-EqB in another tray, and rock for an additional 15 min.
3. Finally, equilibrate in freshly made IAA-EqB for 15 min; drain each strip briefly on its side, then let lay flat, plastic side down on a piece of kimwipe at room temperature until all other strips are ready. *Take care during the IAA-EqB step to protect the strips from light by wrapping the tray with aluminum foil.* During this IAA-EqB equilibration, preparations can be made for the second dimension electrophoresis step.

Second-Dimension Electrophoresis (SDS-PAGE)

1. Second-dimension electrophoresis is carried out using Criterion gels (BioRad). A 10% polyacrylamide gel gives good overall resolution of complex proteomes. Once proteins of interest have been identified through an initial round of gels, a different % acrylamide may be used for the second dimension to optimize separation in a particular molecular weight region. These gels come in disposable plastic cassettes, ready to use. Each has an upper buffer chamber that requires about 50 mL of 1X SDS running buffer. A plastic strip along the bottom of the gel keeps the gel sealed with buffer during shipment and storage; this must be broken off just prior to loading and electrophoresis to allow contact of the gel bottom with the electrophoresis buffer and current.
2. Rinse the top of the Criterion gel with 1X SDS-running buffer; then drain most of the liquid by tipping the gel over a sink, and then blotting one side of the gel upper buffer chamber with Whatman filter paper.
3. Support the Criterion gels vertically in a stable test tube rack, and slip each equilibrated IPG strip onto the top of a Criterion gel, making sure that its edge abuts the gel all along the top of the gel, with no air bubbles or discontinuities between the SDS

gel and the strip. At the same time, make sure the strip is level on top of the gel; do not force part of it into the gel, or between the gel and plate, this will result in a distorted 2D pattern of protein spots.

4. Once the strips are in place, load 5 μL of prestained molecular weight standards in the well on one side. This is optional; any given sample contains its own proteins whose molecular weights can be determined. If repeated gels are run of similar samples, running molecular weight standards is not necessary.
5. The bottom chamber should be prefilled with 1X SDS-running buffer *without SDS*. Some prefer for simplicity to use the same running buffer containing SDS in both chambers; we prefer to have the 10x stock without SDS, and add SDS (10 mL per 2 L) to the upper chamber only. This minimizes having 10X running buffer stock containing SDS sitting around at room temperature, which actually has been known to harbor microorganisms if not used sufficiently quickly. The volume of 1X running buffer required for a full tank of 12 gels is about 5 L.
6. Once the bottom chamber is prefilled with the running buffer, start cooling it to 18°C with water using the tubing system that accompanies the tank.
7. Set the Criterion gels into the slots in the Protean II dodeca tank, and fill each upper buffer chamber with 1X SDS-running buffer (with SDS). Check to make sure all strips are still in position and that all upper chambers contain buffer.
8. Electrophoresis is carried out at 200 V constant voltage until the dye migrates to the bottom of the gels, usually just over 1 h.
9. After SDS-PAGE, the gels can be Western blotted, or fixed prior to staining.

Staining of 2D Gels With the Fluorescent Sypro Ruby Stain

1. Wearing gloves prewetted with water and then gel running buffer, carefully "crack" open the gel cassettes and remove the gels. Handle the gels by the bottoms only.
2. Put into Sypro Ruby fix solution (40% methanol/10% acetic acid) and agitate gently on a rotary shaker for at least one hour; overnight is best. This step is the best place to leave the gel during the staining process. It can be left over the weekend if need be.
3. After fixing, discard the fixative, and immediately add Sypro Ruby gel stain (undiluted out of the container from the vendor). Agitate

gently overnight; protect from light—cover the entire shaking platform if necessary with aluminum foil.

4. The next morning, discard the Sypro Ruby stain, and put in destain solution (10% methanol/10% acetic acid). Continue to protect from light.
5. Change the destain solution at one hour, and change two more times, every hour.
6. Acquire the image of the gel after one hour in the third destain.
7. Once the gels have been stained, destained, and the images acquired, store them in destain solution in seal-a-meal heavy duty bags (protected from light) in the cold. Label so that if you go to pick spots from them 6 mo later, you will be able to recognize a dataset.

2D Gel Image Acquisition

1. Images of the Sypro Ruby stained gels are acquired with the ProXpress imaging system; this is a charge-coupled device (CCD) camera-based imaging system that records images of a 2D gel in rectangular segments, then digitally "*stitches*" these back together.
2. Prepare the ProXpress by wiping the platen with a clean Kimwipe that has been lightly wet with 70% ethanol.
3. Squirt a little water on the platen so the gels are not laid on a bone dry surface, which can "catch" the gels and tear them.
4. Make sure that the gels are free of visible dust or particles.
5. Lift the gels by their bottoms near the corners, and lay them so that the top of the gel lays down first. After a while, one can "feel" what is the best positioning to prevent tearing the gels.
6. For the Criterion gels, three can be scanned simultaneously if two are positioned side by side and the third is laid below them. Images can be rotated once captured, but it is easier if they are squared with respect to the imaging platform, and with respect to each other.
7. For Sypro Ruby, select the excitation and emission wavelengths of 480 nm and 620 nm, respectively.
8. Set the resolution of the images at 100 μm.
9. Choose exposure times that utilize as much of the dynamic range of the imaging system; in other words, time it so that the intensity of the most abundant protein spots approaches, but not does exceed 80–85% of the dynamic range of the imager. The ProXpress generates 16-bit images. The exposure time is optimized to give a

pixel intensity between 50,000 and 55,000 out of 65,536 for the highest intensity spots.

10. Once the images are captured, put the gels back in the destain solution, making sure not to get the gels confused with each other, until the analysis is completed.

Software-Assisted Image Analysis

1. A note about this process: a full description of how to implement 2D gel software- assisted image analysis would take over this entire chapter. It is not expected that a single person will implement all aspects of 2D gel proteomics. More and more, the different aspects are being implemented by "facilities," where investigators may learn "about" 2D gel image analysis, but not actually do it themselves. Thus, this section is written so that the reader can understand the essential features of 2D gel image analysis, but not in such detail as to be able to implement it without going through intensive training for a particular software.
2. Progenesis Discovery image analysis software has an "*automatic analysis*" wizard tool, which allows the user to set up the experiment(s) and let the software implement a first pass analysis without user intervention. Setting up the experiment is relatively quick and easy to do. The automated analysis can take anywhere from one hour to overnight, depending on the size of the 2D gel dataset and the capacity of the computer system; the default procedure is that one starts the automated analysis at the end of the day, for overnight processing.
3. The automatic analysis wizard contains many categories of options that allow for optimized analysis. The options include defining the area of interest in the gel image (the rectangular area in the gel images within which spot detection will be implemented), incorporating different filters that determine how spots will be detected (for example spots above a certain threshold of pixels), setting up groupings of average gels (for biological or technical replicates), and specifying the methods of background subtraction and normalization.
4. After the automated analysis is complete, the data must be manually verified. This involves a visual inspection of the detected spots within the gel images, and may involve "fixing" errors in spot boundaries and spot matches between gels. At this stage of the analysis, image artifacts such as stain speckles, bubbles in the solution on top of the gel, and the dye front that was detected as

spots are dealt with. A fundamental step is the correction of spots that were mismatched between gels, due to gel-to-gel variation, or possibly to sample variation. After these corrections have been made, the software recalculates the background and normalizations. Progenesis also implements what is called "intelligent noise correction algorithm," or INCA, where spikes of stain within or at the edge of a protein spot are deleted, if the slope of the stained spot is such that it is likely a crystal of stain, versus a protein spot that was stained. A stain crystal will have a very steep slope and a flat "top" at the peak of the spot, whereas a protein spot will have a more gradual slope and be rounded at the top.

5. The spot detection data set consisting of spot numbers, *x* and *y* coordinates, nor-malized spot volumes, and INCA normalized volumes is then exported in an Excel spreadsheet for further statistical analysis.
6. Irrespective of which software is used to analyze the 2D gel images, the goals can be summarized as follows:
   (a) Detect as many spots as possible;
   (b) Determine that identical spots are detected as such from gel to gel;
   (c) Determine which spots are NOT on every gel; all information is useful;
   (d) Delete from the analysis nonprotein entities detected as spots such as crystallized stain, dust, air bubbles;
   (e) Ensure that the analysis generates numeric data in Excel format that can be downloaded for statistical analysis.

Statistical Analysis to Identify Significant Spots,and Spots that Discriminate between the Experimental Groups

1. There are different categories of information in 2D gels, and different investigators will have qualitatively different types of goals for the gels they have run. But without doubt, the maximum information from a set of 2D gel images is extracted only after subjecting the image analysis data to systematic and robust statistical analysis. In this section, as in the image analysis section, users are not expected to implement sophisticated statistical algorithms themselves. *How* each of the statistical algorithms is carried out is not detailed in this section. Rather, we discuss the rationale for the different kinds of statistical manipulations. It

should be understood that robust statistical procedures for high dimensional analysis in 2D gel proteomics are evolving, requiring the input of the trained statistician to extract the full information from the 2D gel images. The algorithms and rationales are discussed at greater length elsewhere.

2. Transformation of the data into symmetrical distributions: statistical analysis determines whether an observed difference in abundance or position of a gel spot between gels in different groups is due to the differences in treatment or whether it is due to chance. In order to test if the mean intensity of a protein is different in the two groups, one applies a t-test. But before that, one plots all the spot intensities and looks at the plot. The t-test is most reliable if the plot of the intensities has a symmetric distribution, i.e., if there is roughly the same number of spots with intensities higher than the mean, as there are intensities lower than the mean. If there seem to be a small number of really high-intensity spots and a large number of low-intensity spots, it is a good idea to do log transformations of the intensities and use these to test the differences in the means rather than the raw intensities.
3. The t-test is a univariate test, in that it looks at one protein at a time. However, most biological proteins do not function independently of each other, thus they probably react in concert to various treatments. In order to allow for this biological reality in the statistical tests, one uses "*multivariate*" tests, i.e., one tests for differences that distinguish groups of proteins instead of between single proteins. A t-test or other similar univariate test, however, is often used to reduce the number of proteins one uses in a multivariate test.
4. A multivariate test can identify a group of proteins that can differentiate a gel in one group from a gel in the other group. These tests are usually called classifiers. *Discriminant Analysis* (DA) is a classifier. In this test, one measures the distance of the intensity of the chosen group of gel spots in a gel from the centre (mean) intensity for those spots in each group of gels. Gels within a predetermined threshold are assigned to the group they are closest to. This is repeated for each gel until all of the gels are classified into a group. Because the true group of a gel is known, one can then determine the quality of the DA by looking at the true-positives vs false-positives. To give an extreme example, if we had a set of protein spots that were present in all treatment gels

and absent from all the control gels, then the summed intensity of these protein spots in the control gels would be zero. The algorithm looks for these spots (by their spot numbers previously determined by the image analysis software) in a list of spots for a gel; if it finds that the intensities of these spots on that gel sum up to zero, it will classify that gel as a control gel. The process will work in the opposite way for a gel in which the intensities of the spots add up to a number that is close to the mean intensity of those spots in the treatment gels. However, if there is a treatment gel in which these spots exist but are of very low intensity, so that their sum is closer to zero than to the mean value for these spots in the other treatment gels, then the gel will be falsely classified as a control gel. It is to be noted that the actual summed quantities are not just the summation of the intensities, but rather functions of the intensities and correlations between the selected protein spots. It is the contribution of the correlations that makes this a true "*multivariate*" analysis. In our study of the actions of grape seed in rat brain, DA identified seven gel spots that discriminated 100% between the two experimental groups. In other words, although a number of spots were determined to be significantly different in their intensities between the two groups, the sum of the variances of these seven between the two groups was all that were required to distinguish them.

5. Principal Components Analysis is a computational technique that graphically displays the differences in the groups in an experiment (if a difference exists). In a conventional 2D gel, the x coordinate is a measure of the isoelectric point or pI of the protein, and the y coordinate is a measure of the mass of a protein. The plane defined by the two axes is the 2D gel. In a principal components plot, the x and y axes represent in part the sums of the gel spot intensities in a gel. Yet these are not simple additive quantities, but combinations of gel spot intensities; the first component is the combination of spot intensities that has the largest variance and the last component represents the combination that has the smallest variance. The software calculates the values of these various combinations for each gel. The gels are then plotted in the plane defined by these components. The idea is that if the variability in the gels is related to the differences in treatment, i.e., if the spot intensities in one set of gels vary a great deal from the spot intensities in the control gels, then they will form clusters in

different regions of the plane defined by the components. PCA can be used before discriminant analysis, to see if it is likely that discriminant analysis will provide a good separation of the two groups, or after discriminant analysis, to demonstrate visually the separation of the two groups by the proteins identified in the DA.

Robotic Gel Spot Excision, or "Picking"

1. Gel spots of interest (indicated by the image and statistical analysis) are excised using the robotic ProPic. Sypro Ruby-stained spots require ultraviolet (UV) illumination for detection. Most spot pickers are electronically driven in the x and y dimensions by spot coordinates generated within the image analysis software. The spot picker picks spots in numerical sequential order by spot number. The excised plugs are then deposited into the wells of a 96-well plate for destaining, prior to incubation with trypsin or other protease.
2. Although Progenesis and the ProPic can align the gel with the previously obtained image obtained during image analysis by aligning a set of "*landmark proteins*" around the area containing the spots of interest (triangulation) with the original image, this option is not, in fact, used routinely with the Criterion gels—the latter are too small for the triangulation to be used effectively; there is too little room for error. With the Criterion gels, less error is introduced if one "picks" based on direct visual assessment of which spot in the previously analyzed image aligns with which spot in the re-visualized gel in the spot picker. In this mode, click on the spot to be picked.

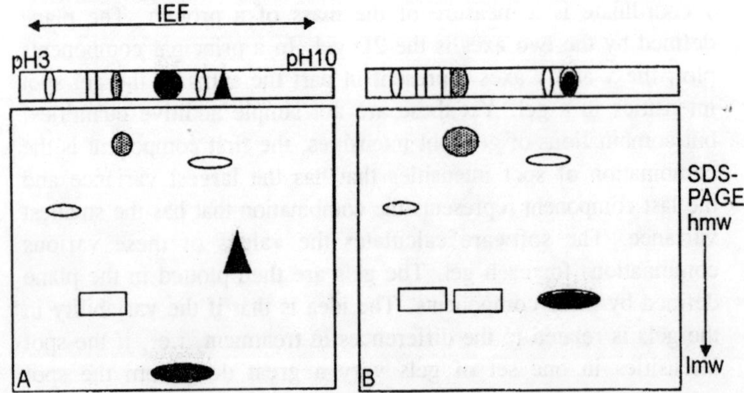

Fig. 15.1. Two-dimensional gel electrophoretic separation of polypeptides:

In-Gel Digestion

The protocols described here for in-gel digestion and for peptide mass fingerprint analysis are based closely on procedures developed at University of California San Francisco National Mass Spectrometry Resource, and as modified by Speicher and colleagues.

1. Each gel plug is covered with 50 μL Milli-Q water, and agitated in a rotating wheel for 10 min, after which the water is discarded. This step is repeated once.
2. Both Coomassie blue-stained and Sypro Ruby-stained plugs are processed similarly: two 30 min incubations in 100 μL of 50% acetonitrile/50 m*M* NH_4HCO_3. This step is repeated if stain persists in the gel pieces.
3. The destained gel plugs are dried passively in their tubes in an incubator at 37°C until no moisture is detected in the tubes. At this point, the gel plugs should be opaque.
4. A trypsin stock (12.5 μg/mL) solution is diluted 1:1 with 25 m*M* NH_4HCO_3. A 10 to 20 μL aliquot of this diluted trypsin is then added to each dried gel plug. Allow 10–15 min for the gel to reswell, during which time the trypsin is absorbed into the gel. Add additional 25 m*M* NH_4HCO_3 to just cover the gel pieces. These are incubated overnight at 37°C.
5. After overnight incubation, the trypsin digestion is stopped by adding 50 μL of 5% (v/v) formic acid, after which the tubes are sonicated in a water bath sonicator for 30 min. The supernatant is removed and placed in a new 0.5 mL Eppendorf tube.
6. To recover remaining peptides in the gel pieces, 50 μL of a 50% acetonitrile/5% formic acid is added, and the tubes are sonicated for 30 min in a bath sonicator. The extracted solutions are pooled at this point in the new centrifuge tube, dried down to 10–20 μL by Speedvac evaporation, and desalted by passage through C_{18} ZipTips. Of the resulting volume, 0.5 μL is committed to matrix-assisted laser desorption/ionization (MALDI)-time-of-flight (TOF) MS analysis. The remainder of the sample may be processed by liquid chromatography (LC)-electrospray ionization (ESI)-tandem MS to confirm by sequence analysis the protein identity suggested by the MALDI-TOF MS analysis.

Peptide Mass Fingerprint Analysis Using MALDI-TOF MS

Many institutions provide MS analyses in the form of shared/core facilities to investigators, where an "operator" carries out the MALDI-

TOF MS of peptide mixtures from gel spots, and hands the investigator the data. The rationale for carrying out the analysis in such a facility is to reduce the contamination background—the operators are experienced in maintaining a clean environment and the use of the robotic equipment is a distinct advantage. Nonetheless, some investigators may prepare samples for MALDI-TOF MS.

1. The sample containing peptides is mixed with a saturated solution of the "matrix" in 50% aqueous acetonitrile (usually in a 1:1 mixture, followed by further dilution with the matrix solution). The mixed sample (0.5–1 μL) is spotted onto the surface of the MALDI target plate (typically stainless steel and some cases coated with gold) and allowed to air-dry. The plate is typically marked with a numbered 10 × 10 array of spots. The position where each sample is spotted must be noted. The most commonly used matrix for proteolytic peptides is α-cyano-4-hydroxycinnamic acid. Once loaded with the samples to be analyzed, the plate is introduced into the MALDI-TOF MS.
2. Figure 15.2 is a "raw" MS spectrum from a MALDI-TOF analysis of a gel spot digested with trypsin. Each peak represent a different peptide fragment generated from the digestion process. *Peak A* is a peptide with a *m/z* (mass/charge) value of 2012.99 resulting from the digest of the desired protein. *Peak B* is a peptide from the proteolysis of trypsin. Trypsin can cleave itself (also known as *autocatalysis*). The mass spectrum can be calibrated using these internal autocatalytic peaks for better mass accuracy when searching for the identity of protein.
3. In order to identify the protein successfully, the spectrum must be simplified by a process called *de-isotoping*. Each peak is actually a series of clustered isotope peaks that occur because of natural abundance of carbon-13 (approx 1% of all carbon atoms). The abundance of carbon-13 isotope peak increases with each additional carbon atom. In the case of 15-residue peptide, the number of carbon atoms is approx 100. For such a peptide, the intensity of carbon-13 isotope peak will approximate or exceed that of the all carbon-12 peak. **Figure** demonstrates the phenomenon of isotopic distribution in the spectrum. In this case, the intensity of the *m/z* 1367.82 peak containing one carbon-13 atom is 90% of that of the *m/z* 1366.82 peak (all carbon-12). In order to get an accurate identification with a database, the spectrum must be simplified to the carbon-12 only peaks. This process is usually carried out with software such as *Voyager Explorer* or other similar programs.

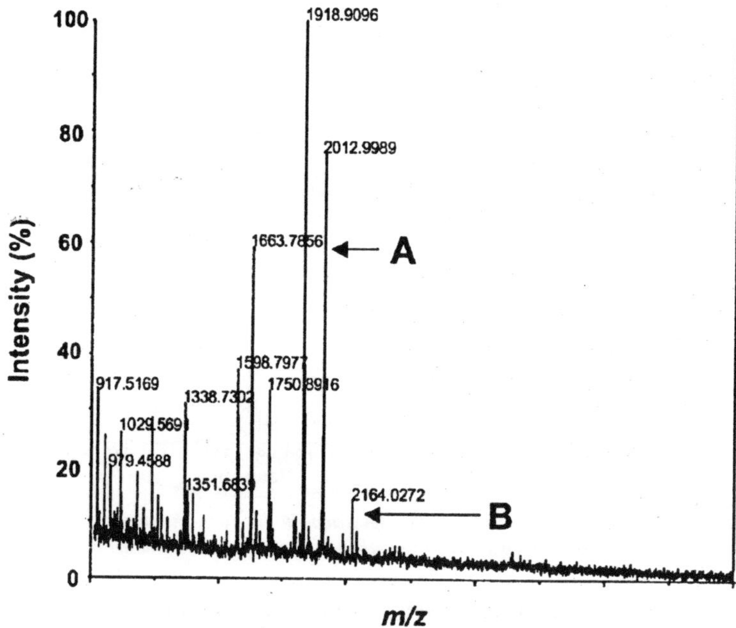

Fig. 15.2. "Raw" mass spectrometry (MS) spectrum from a matrix-assisted laser desorption/ionization (MALDI)-time-of-flight (TOF) MS analysis of a typical gel spot digested with trypsin.

4. Once the mass spectrum is properly edited, a database search is carried out using one of many different programs. In one example, the carbon-12 peak data minus the autocatalytic fragments can be put into MASCOT. Typically, the MASCOT search is conducted allowing for one missed tryptic cleavage, methionine oxidation, and a mass accuracy error of less than 0.1 Da. Other proteases can be used instead of trypsin—these include Lys-C (cleaves at lysine residues), Arg-C (cleaves at Arg residues, chymotrypsin (cleaves at phenylalanine, tyrosine and tryptophan residues), and Glu-C (cleaves at aspartate and glutamate residues). Additional fixed molecular weight changes due to alkylation of sulfhydryl groups of cysteine residues, or variable changes, for example, due to phosphorylation of Ser, Thr, and Tyr residues, can be applied to the search analysis. MASCOT returns information on the most likely matches to the ion masses from the spectrum. This consists of the protein name, a molecular weight search (MOWSE) score, the protein Accession Number (peculiar to the database chosen for the matching), and molecular weight of the intact protein. The

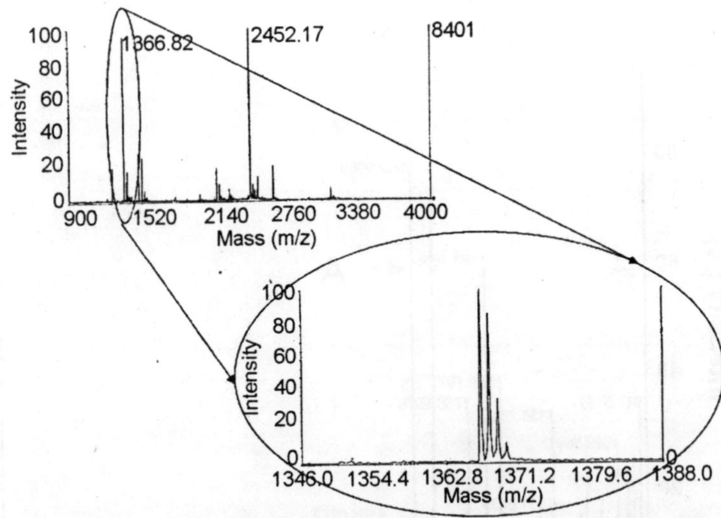

Fig. 15.3. Isotopic distribution in a mass spectrum. In this case, the intensity of the m/z 1367.82 peak containing one carbon-13 atom is close to that the m/z 1366.82 peak.

score is based on the several factors including mass accuracy of the peptides matched in the mass spectrum, the number of peptides matched, and sequence coverage of the protein. In this case, the MASCOT algorithm determined that the best fit of the peptide mass data was to a "Lipoprotein" from *Mycoplasma pulmonis* with a MOWSE score of 134. This MOWSE score fell outside of the shaded box on the MASCOT file; hence, the protein identification is considered "significant." MASCOT also provides MOWSE scores for proteins with lower degrees of significance. In the case shown previously, other proteins were inside the green shaded box and therefore were non-significant matches. Additional information is linked to the accession number, including a list of theoretical tryptic fragments, color-coded to show those discovered in the observed mass spectrum.

Protein Identity Confirmation by LC-ESI-MS/MS

Whereas MALDI-TOF MS offers putative identification of proteins, LC-ESI-MS/MS (liquid chromatography-electro spray ionization-tandem mass spectrometry) can confirm the identification by analysis of ions derived by collisional dissociation of the parent peptide. This type of MS provides information about a peptide's amino acid sequence, complementing the mass information generated by MALDI-TOF MS analysis. It is assumed that the investigator has access to an appropriate mass spectrometer operated by a trained operator. This section focuses on understanding and interpreting the spectra obtained from an LC-ESI-MS/MS experiment, rather than on how to operate the instrumentation.

1. Fragmentation of a selected peptide occurs in the mass spectrometer. This fragmentation called collision-induced dissociation (CID) occurs by accelerating the peptide ion into neutral argon gas. A principal site of cleavage is at the amide bond generating "B" and "Y" ions. B ions originate from the N terminus and consist of the amino acid residues plus a hydrogen. Y ions originate from the C terminus and consist of the amino acid residues plus hydrogen and one H_2O.
2. Figure 15.4 shows an example of LC-ESI-MS/MS data, where the sequence has been determined, based on the B and Y ions identified. As indicated in the figure legend, the known masses of amino acid residues are used to identify the amino acids in the sequence of B and Y ions.
3. Once peptide sequence is obtained, the sequence is entered at the free website. This website searches the NCBI nonredundant (nr) database and returns the polypeptide(s) that contains a peptide of this sequence. In cases dealing with protein families such as kinases, more than one peptide sequence may be needed to positively identify the specific protein, due to sharing of homologous domains. If the sequence is "novel," i.e., not in the database, then it is important to reinspect the data for any errors in interpretation. It should be noted that more and more genomes (and hence proteomes) are being described. Nonetheless, proteins not readily identified as

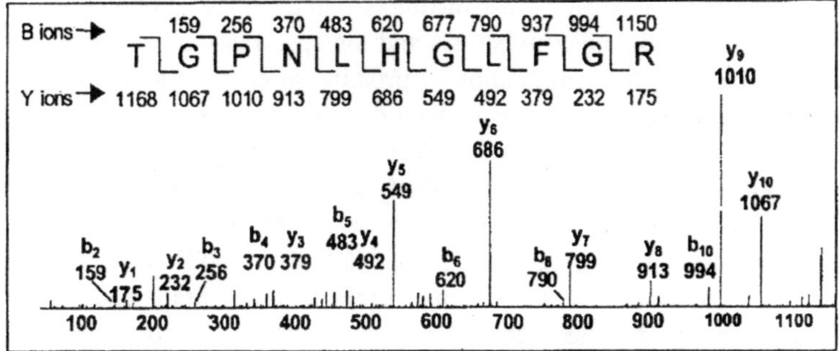

Fig. 15.4. Liquid chromatography (LC)-electrospray ionization (ESI)-tandem mass spectrometry (MS/MS) analysis to determine the sequence of a peptide ion.

being derived from the DNA coding strand have been reported, raising the possibility that consideration of novel mechanisms for the translation of DNA sequence information may become necessary. Regardless, these will, no doubt, rapidly appear in the proteomics databases.

4. A popular alternative strategy for identifying the sequence of a peptide is to use a search engine such a MASCOT or SEQUEST. In the case of the latter, the molecular weight of the parent peptide ion is used to first select all expected peptides within 1 Da from the database being used. The peptides in the subdatabase are individually subjected to theoretical fragmentation and are compared to the ions in the tandem mass spectrum of the unknown peptide. The peptide in the subdatabase having the closest fit to the unknown peptide is chosen as the "match", along with a score. This method allows for the "identification" of up to 50,000 peptides per sample. However, experiments using a control database revealed that a large proportion of the matches were false-positives. Improvements in the scoring algorithms for the analysis of known peptide tandem mass spectra, as well as *de novo* procedures, are sorely needed.

Study of Age-Related Protein Oxidations by 2D Gel Proteomics Technology

Protein oxidations resulting from accumulated reactive oxygen species are widely considered to be part of the disease process in diverse age-related conditions including rheumatoid arthritis, cataractogenesis, diabetes, atherosclerosis, and Alzheimer's disease. These modifications can impact protein function in various ways by modifying amino acids essential for activity, causing changes in tertiary structure, or by abnormal cross-linking. Combining the resolving power of two dimensional electrophoresis and the specificity of Western blot analysis can enable identification of proteins susceptible to age-related oxidations, as well as analysis of how different treatments affect them. The protocol described here uses established Western blot methodology using a commercially available antibody that recognizes a dinitrophenyl (DNP) moiety covalently attached to an amino acid resulting from derivatization with dinitrophenylhydrazine (DNPH). DNPH derivatization, was used to show that aged human brain and AD brain had higher levels of oxidized proteins than young human brain. Although specific proteins (glutamine synthetase and creatine kinase) were shown to have lowered activity in aged brain, neither was specifically examined for

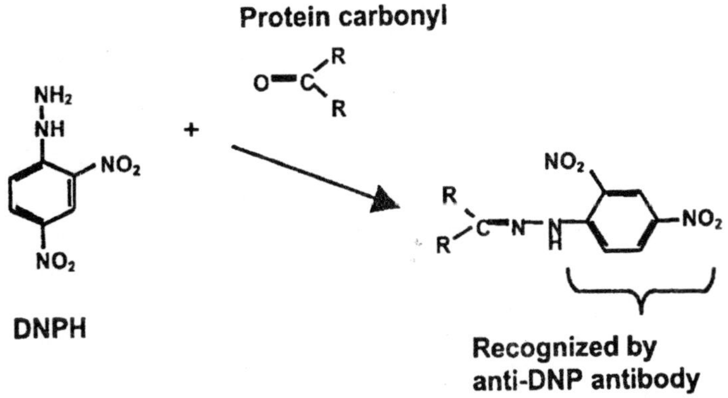

Fig. 15.5. Reaction scheme for derivatization of protein carbonyls with dinitrophenyl-hydrazine.

extent of oxidation, nor was a global analysis done to determine what other proteins might be oxidized; however, these results suggested for the first time that increased protein oxidations might be involved in the pathology of aging, and in AD. Using the DNPH derivatization protocol described here, 2D gels and replicate Western blots of AD brain showed several protein spots that were different in oxidation level relative to age-matched control brain, but preliminary quantitation suggested that oxidation was not necessarily increased in AD; moreover the identities of the spots were not revealed. Additional studies using the same DNP-derivatization assessed that total age-related protein oxidations were increased in aged wild-type, or aged transgenic mouse models of dementia; however, there was no attempt to examine specific proteins. Our own studies showed that recombinant human *creatine kinase brain* isoform (CKB) could be oxidized readily by the reactive aldehyde 4-hydroxynonenal (4HNE) on multiple sites, and that such oxidations abolished enzyme activity, thus extending the observations of Aksenov et al. (2000), who showed that CKB was oxidized at higher levels in AD brain and suggested that this was the basis for the reduced enzyme activity. In no case, however, has the oxidation of any protein been clearly shown to be causal to the aging process, or to the pathogenesis of AD. Clearly, much work must be done in this area.

1. Reagents for Western blot analysis of protein carbonyls:
   (a) OxyBlot Protein Oxidation Detection Kit
   (b) 18% SDS stock solution
   (c) Immobilon-FL PVDF membrane
   (d) Odyssey Blocking Buffer

(e) Anti-DNP Primary Antibody

(f) Goat anti-rabbit secondary antibody conjugated with Alexa fluor 680

2. Dinitrophenylhydrazine (DNPH) derivatization of protein carbonyls:

   (a) Adjust 10 μL of IEF buffer containing 30 μg of protein to 6% SDS by adding 5 μL of an 18% stock solution.

   (b) Add 10 μL DNPH (dinitrophenylhydrazine) derivatization solution from the OxyBlot Protein Oxidation Detection Kit; incubate at room temperature for 20 min. Protect from light by wrapping the whole rack of tubes with aluminum foil, or by putting the rack in a drawer.

   (c) Terminate the reaction by the addition of 7.5 μL of neutralization buffer provided in the kit.

3. Methanol-chloroform precipitation of proteins. After quenching the DNPH-derivatization, precipitate the proteins by chloroform methanol extraction as follows:

   (a) add 4X initial sample volume of methanol, vortex;

   (b) Add 1X initial sample volume of chloroform, vortex;

   (c) Add 3X initial sample volume of water, vortex;

   (d) Centrifuge at 10,000 rpm for 2 min, and remove the supernatant;

   (e) Add another 3X initial sample volume of methanol, vortex;

   (f) Do a final centrifugation at 10,000 rpm for 2 min;

   (g) Remove the supernatant, and resuspend the protein pellet in 200 μL IEF lysis buffer;

   (h) Determine the protein concentration using the Bradford.

4. 2D Western blot analysis for protein carbonyls:

   (a) Carry out 2D electrophoresis. Run replicate gels for each sample; fix and stain one with GelCode Blue Stain Reagent according to the manufacturer's directions; process the other unfixed through Western blot analysis.

   (b) After the SDS-PAGE, equilibrate the unfixed gel in transfer buffer for at least 15 min. The purpose of this is to allow the free, non-protein-bound SDS to diffuse out of the gel; discard the buffer after this step.

   (c) While the gel is equilibrating, activate the FL-PVDF membrane by soaking in 100% methanol for 5 min—this is crucial for hydrating the membrane;

(d) Equilibrate the membrane in transfer buffer for 15 min.

(e) Assemble the gel in the Western blot cassette with the FL-PVDF membrane (precut to the same size as the gel). Make sure there are no air bubbles between the gel and the membrane. This can be facilitated by laying the membrane down first, then the gel. The complete sandwich inside the cassette should consist of sponge, Whatman filter paper, membrane, gel, Whatman filter paper, sponge. In the completed sandwich, the gel should be snug against the membrane, otherwise slippage will occur and the transfer will not be a mirror image of the 2D gel.

(f) Insert the cassette in the transfer tank so that the gel is closest to the anode, or so that the direction of transfer is gel to membrane.

(g) Electroblot overnight at 130 mA, with the transfer tank sitting in a dishpan filled with ice, and the buffer in the tank being stirred with a stirring bar.

(h) After transfer, the blot is carefully removed from the cassette, and processed through antibody incubation.

(i) Block for 1 h at room temperature in Odyssey Blocking Buffer with agitation.

(j) Incubate in anti-DNP 1:2500 in Odyssey Blocking Buffer, 1 h.

(k) Wash the blot with six 10 min changes in TBST.

(l) Incubate in Alexa Fluor 680 goat-anti-rabbit immunoglobulin (Ig)G antibody 1:5000 in Odyssey Blocking Buffer, 1 h *in the dark* (Either wrap the box in foil, or use light-tight boxes).

(m) Wash overnight in TBST, at 4°C.

(n) Acquire the image of the blot on the LiCor Odyssey Scanner following manufacturer's instructions.

(o) Determine the spots from the stained gel to be processed for MS by a combination of visual inspection of the stained gel image and the Western blot image, and/or in conjuction with image analysis using Progenesis.

5. MALDI-TOF MS. Gel spots are excised and processed through MALDI-TOF MS.
6. Identification of protein modifications by LC-ESI-MS/MS: post-translational modifications can be indicated and even quantified by Western blot, However, the latter does not identify the site of modification on the protein. Identifying the exact chemistry of the

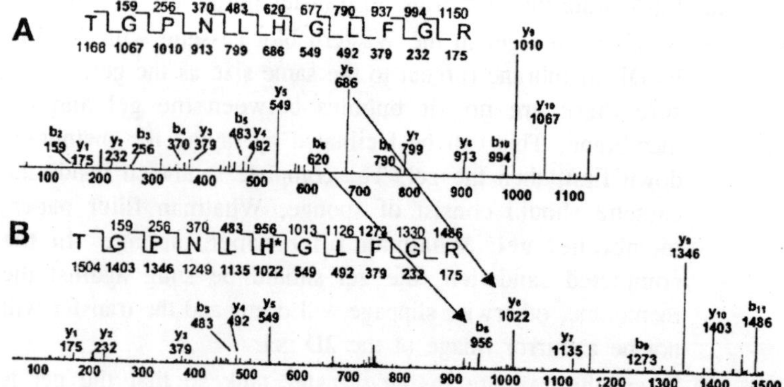

Fig. 15.6. Liquid chromatography (LC)-electrospray ionization (ESI)-tandem mass spectrometry (MS/MS) identification of peptides oxidized by 4-hydroxy-2-nonenal (HNE).

modification on the protein, as well as mapping it to a specific amino acid residue, not only confirms the modification indicated by the Western blot, but also gives insight into functional aspects. Pinpointing the site of modification is often a difficult task. Having a nonmodified protein as a control to compare with a modified protein is essential in dissecting the spectra and identifying the modified amino acid.

Mass spectrum A is the same as given above. This is an unmodified peptide. Mass spectrum B, however, is the same peptide that was oxidized by incubation with the reactive aldehyde 4-hydroxy-2-nonenal (HNE). This oxidized protein was then derivatized with DNPH as described above, followed by an overnight digestion with trypsin as described above. The expected mass shift due to the addition of one HNE adduct is 156. The expected mass shift of one HNE adduct derivatized with DNPH is 336. In B, all of the B ions from the histidine residue (H) to the C terminus have a mass changed by the addition of 336 and all of the Y ions from H to the N terminus also have this addition. These data confirm that the HNE adduct is on the histidine and is a Michael adduct.

2D Blue Native electrophoresis for Analysis of Protein Complexes

The "blue native gel" enables analysis of native (nondissociated) protein– protein interactions, particularly those involving hydrophobic proteins in membrane environments, by releasing the complexes from the membrane environment with a mild detergent, without dissociating the complexes. Originally developed to analyze mitochondrial oxidative complexes in the mitochondrial inner membrane, the method has begun to be used to study nonmitochondrial protein complexes in the cytosol or in the plasma membrane. To prevent re-aggregation of hydrophobic membrane proteins, the solubilization buffer contains a high concentration of aminocaproic acid; being a zwitterion, free aminocaproic acid does not migrate in the gel, thus avoiding issues of depletion from the proteins during electrophoresis. Finally, the solubilization is carried out in a low concentration of Coomassie Blue G-250, which associates with the proteins and, in doing so, results in the important consequence of the proteins all having the same type of charge, so they all migrate in the same direction (toward the anode at neutral pH). Thus a mixture of protein complexes of different native charges can be loaded at the top of a native gel, and be resolved according to differences in apparent native molecular weight. Once a BN gel has been run, a lane from this gel can be equilibrated in SDS sample buffer, laid on top of a SDS gel, and conventional SDS-PAGE carried out, resulting in a 2D separation of a category of proteins that might otherwise be difficult to study by other electrophoretic means. The second dimension can be stained and assessed visually as to differences, to determine "bands" to excise for MS analysis.

1. Chemicals and stock solutions:
   (a) Coomassie Brilliant Blue G-250, powder, room temperature.
   (b) Lauryl-maltoside (dodecyl-β-D-maltoside) 10% w/v, 0.1 g in 1 mL water, –20°C.
   (c) Make the stock solutions for the BN electrophoresis buffers.
   (d) High-molecular-weight standards.
   (e) Resuspend contents of one vial in 200 μL of extraction buffer, then add 25 μL of lauryl-maltoside and let sit on ice for 20 min.
   (f) Add 12.5 μL of Coomassie Blue (containing the aminocaproic acid) to the standards mixture just before loading onto the gel. Store remainder at –20°C.
2. First dimension BN gel:
   (a) Assemble the gel cassette with cleaned glass plates, and 0.75 mm spacers.
   (b) Make 10 mL of each % acrylamide in 10 mL graduated cylinders. Bring to volume with cold water, seal with parafilm, and mix gently by inversion several times, then let sit in an ice bucket until ready to pour the gels. It is important to keep cold, otherwise the acrylamide will begin to polymerize before the gels are poured.

(c) With the connection between the chambers closed, fill the RIGHT side (closest to the gel assembly) first with 4.2 mL of the 12% acrylamide.

(d) Open the connection to allow the 12% acrylamide to fill the connection so that you can just see it come into the other chamber. This prevents air bubbles from being trapped in the connection that will prevent flow between the two chambers. It is preferable to fill with the denser solution first, since it will tend to fill the connection anyway, due to it being more dense. Fill the chamber further away from the exit port with the 5% acrylamide; its level should just barely exceed that of the 12%.

(e) After both chambers have been filled, add the APS and TEMED to both, swirling with plastic pipets to ensure immediate and thorough mixing.

(f) Start the stirring bar gently stirring in the chamber containing the higher percent acrylamide; it should maintain a vortex, but not so deep that air bubbles are introduced.

(g) As rapidly as possible after adding the APS and TEMED, smoothly open the stopcock on the chamber closest to the gel assembly, and allow the solution to flow into the gel assembly cassette through a pipet tip attached to the outlet tubing; for stability, this tip should be taped to the gel plate assembly, between the plates. Lower the stirring bar speed to minimize bubbles entering the gel when the level nears the bottom. You should be able to see the mixing of the different acrylamide concentrations in the right chamber during the pouring of the gradient. If you see mixing in the LEFT side of the gradient, it means you have backflow, and the gel probably will have non-reproducible discontinuities in the gradient. When in doubt, let the gel polymerize, then discard that gel, and pour another one. Tilt the entire mixer to allow for last bit of gel solution to enter the gel assembly.

(h) When all the gel solution has poured out of the gradient maker, move the outlet tip away from the gel, tape it to the edge of a waste beaker, and IMMEDIATELY pour water into both chambers of the gradient maker and let flow into the waste container, to prevent residual acrylamide from polymerizing within the tubing, or drying within the chambers. Drain the chambers by tilting the assembly toward the waste container.

(i) For more reproducible gradients and greater ease of generating gradients, a peristaltic pump should be used. This avoids having to raise the gradient maker above the gel to get the flow started, and enables better control of the flow. Also, rinsing after pouring a gel is MUCH easier with the pump.

(j) After rinsing the gradient maker with water, overlay the gel gently with H_2O-saturated isobutanol. This is lighter than water, and thus disturbs the top of the acrylamide less when it is pipetted on top of the just poured gel. Nonetheless, care should be taken to minimally disturb the gel, keeping the just-poured gradient in mind. Pipet the isobutanol on top of one side of the gel with a one-piece plastic pipet, and let the butanol layer "glide" smoothly across the top of the unpolymerized gel solution, until it reaches the other side. Don't let it "bomb" through the gel; if this happens, probably the gel should be discarded.

(k) To pour the stacker, partially insert the desired "comb" (10 or 15 wells) into the gel assembly above the polymerized separating gel so that the part that will form the top of the stacker between wells is above the plates, and pour the acrylamide solution for the stacker, letting it overflow a bit out of the gel. Then slide the comb down into the gel so that the well depth is maximized, but allowing at least 3–5 mm stacking gel above the separating gel.

(l) The first dimension electrophoresis must be cold, so ensure that the buffers are.

(m) Put the Anode buffer into the gel tank.

(n) Take the gel assembly out of the gel pouring assembly without removing the combs. With a black magic marker, mark the positions of the well bottoms on the gel side facing the outside of the box so they will be visible.

(o) Click the gel into position making the upper buffer chamber, make sure the gasket has sealed, and test the seal by filling the chamber with water with the assembly on a piece of lab matting. If it does not leak, pour the water out.

(p) Place the gel assembly into the gel tank filled with Anode buffer. Fill inner reservoir with Hi Blue Cathode all the way to the top.

(q) At this point, the comb can be gently pulled out. If you look closely, you can see the wells made by the comb.

(r) Using disposable gel-loading pipet tips, load the wells with desired samples prepared appropriately. Up to 30 μL can be accommodated in the wells made by a 10-wellcomb.

(s) Secure the lid on the tank, making sure that the leads on the gel assembly fit into the lid in the correct orientation, red to red and black to black.

(t) Take the tank to the cold room, connect to the power supply, and turn on the power. Run the gel at 40 V for about an hour, to allow the proteins to migrate through the stacker.

(u) Then, aspirate out the inner buffer, and put in Lo-Blue buffer, also cold. Increase the voltage to 100 V and run for about 2 h.

(v) At the end of the run, take the gel assembly out of the tank, dump the Lo-Blue buffer out of the inner reservoir into the sink, open the cassette with a nonsharp wedge (provided by BioRad) and carefully (with gloves on) put the gel in Coomassie Blue gel stain/fix for about 2 h on shaker.

(w) Transfer gels to a new staining dish containing Destain, roll up Kimwipes, and position the rolls at the sides of the dish, away from the gels. These bind Coomassie Blue, and speed up the destaining.

(x) Complete destaining may take all day, but the band pattern should be detectable within a few hours.

3. Second dimension of the BN gel procedure. Tris-Tricine gels are used here, because of their higher resolving power with low molecular polypeptides.

For the second dimension, the BN gel (the first dimension) is NOT stained or fixed; this will precipitate proteins and migration in the second dimension will not occur .

(a) Turn on the heating block to 80°C.

(b) Remove the agarose solution from 4°C storage, and heat in the heating block. About three tubes will be needed for two gels.

(c) Before the BN dimension is done, prepare your second-dimension gels. For analysis of mitochondrial complexes, the most standard gel consists of a 10% acrylamide layer, with an 8% acrylamide layer on top. For other samples, the optimal composition will have to be empirically determined. It may be that a simple 10% layer will suffice.

(d) Assemble the gel cassette as before for the BN dimension. Premark the gel plate to 5.5 cm from the bottom using a ruler; this height ensures maximal separation in the 10% layer. Pour the acrylamide solution into the assembly with a disposable 10 mL pipet smoothly with a motorized dispenser. Overlay the gel with the same H_2O-saturated isobutanol as before for the BN dimension.

(e) When the 10% layer has polymerized, pour the stacker to within 5 mm of the top of the plates; this allows sufficient depth above the gel to slip in the lanes from the BN gel. The specific depth of the 8% layer is not critical, but if running more than one second-dimension gel in the experiment, pour these sections to the same height in all the gels.

(f) Make sure the agarose solution is heated thoroughly so it is a free-flowing liquid. When it goes up and down freely in a plastic one-piece pipet, it is ready.

4. Preparing the BN gel lane for second dimension:

(a) Disassemble the gel cassette after the BN dimension, and discard the stacker, cutting through the gel between the stacker and the separating gel with a razor blade.

(b) Using a one-sided razor blade as a guide, cut a lane from the BN gel just about the same length as the blade, measuring from the top of the gel, after the stacker has been discarded. This avoids the large blue "*blob*" at the lower end.

(c) Cut the BN lane no more than 5 mm wide.

(d) Pipet the hot agarose solution on top of the stacker in the second dimension gel positioned at an angle against a support. Gently but quickly slide one of the two BN lanes into the space between the plates, sliding it all the way to the left.

(e) Insert the second BN lane next to the first, in the same orientation as the first.

(f) Insert a Teflon comb section in the hot agarose at one end to form the well for loading the standards, being careful not to push it below the level of the BN gel pieces.

(g) Finally, dribble denaturant solution on top of the BN gel lanes to ensure denaturation. Let the whole setup sit for 15–30 min.

(h) Start the gel at 30 V for about an hour, or until the samples have migrated out of the BN gel and the narrow dye migration front is visible.

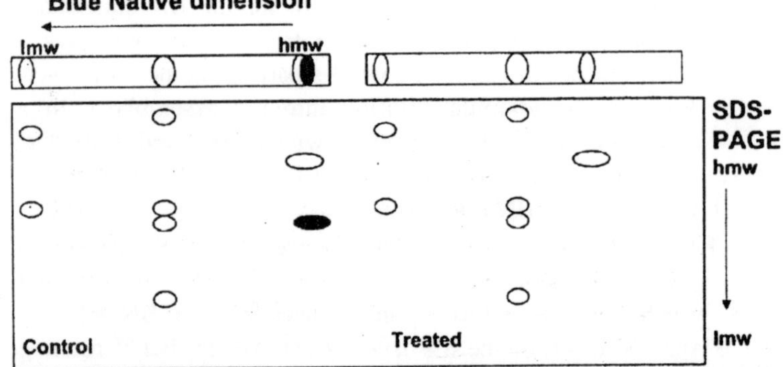

Fig. 15.7. Schematic of two-dimensional blue native electrophoresis.

(i) Increase to 120 V, and run for about 2 h, or until the smallest of the standards is at the bottom of the gel.

(j) Disassemble, and stain the gel as before with Coomassie Brilliant Blue R.

Notes

1. General quality control: considerations: what do we mean by quality control? "Quality" refers to the goodness of the data; quality control then refers to how an experimental protocol is defined to generate as good data as possible. There are several parameters: data are good if they are *reproducible*, if they are *unbiased*, and if they are *meaningful biologically*. These are not specific to aging-related research, but where time points are involved, such as in many experiments addressing aging, attention to quality control is critical to generate meaningful data.
2. Recognizing sources of bias: even before the proteomic analysis is started, bias can be introduced by instrument drift, by the way that samples are collected, by the way the samples are processed, or by the way the samples are stored. If at all possible, the collection of biological samples should be randomized, so that similar samples are not collected all on one day, and another group all on another day. With timed experiments, time points obviously dictate a nonrandom order of sample collection. To minimize bias, make sure that all samples are collected in the same type tube/container, using the same stop solution. It would be better to randomize the order of initiation of the experiment, so that different time points were collected in random sequence vs from shortest incubation time to longest incubation time. If

this cannot be done, then the samples should be randomized following the experiment, so that time of analysis is not a factor. Sample processing and storage become important issues with multiple samples. "*Homogenization*" can mean different things to different technicians. If at all possible, have the same person process all samples in a dataset, using the same type of homogenizers and other equipment. Brains should not be homogenized one day with a glass-glass homogenizer, and the next day with a glass-teflon homogenizing system. Similarly, centrifugation conditions should be exactly the same for all the samples. Room temperature can vary. The centrifuge should be set at 22°C, or 20°C. Often, the exact conditions are not critical, but the fact that all samples are exposed to the same conditions is.

Randomization to minimize bias: bias creeps in at points in the experimental process that we do not think about. In setting up Western blot cassettes in the larger BioRad tanks, up to three cassettes can be run simultaneously. But care should be taken to have replicates of samples in different cassettes, and samples being compared in the *same* cassette, so that if there are differences in transfer due to positioning of the cassettes with regard to the electric current, these differences will affect pairs of samples being compared uniformly. By noting whether the pattern of proteins between replicates transferred in different cassettes is the same, one can determine that the positioning does not affect extent of transfer, or if it does, that all samples to be compared are transferred in the same cassette, but technical replicates can be transferred in different cassettes.

3. Understanding the chemistries used in 2D gel generation: reduction and alkylation prior to IEF may make a difference. The protocol described here for reducing and alkylating the IPG strip just prior to SDS-PAGE has worked well for us and others for numerous kinds of samples. Recently, however, systematic analysis showed that alkylation with IAA in SDS-containing buffer resulted in incomplete alkylation after even one hour, whereas CHAPS inhibited far less, suggesting that alkylation before the IEF step in IEF buffer might be far more effective than during equilibration in SDS-buffer; moreoever IAA was far less effective than the vinylpyridine family. Bai et al. (2005) showed recently that alkylation with vinyl pyridine gave 2D gel patterns with greater resolution in the basic region, presumably reflecting complete

alkylation, and thus less re-formation/formation of disulfide bridges. These results indicate that consideration should be given to the reduction and alkylation conditions described in numerous proteomics publications, including this chapter. In general, the investigator is cautioned to optimize recipes even where they appear to be "standard" protocols; often, procedures become implemented "standard" protocols without careful assessment of whether they have been validated, or whether they are optimal for their particular samples or experiments.

4. To reduce gel to gel variation: for conventional 2DE, we have found that running single % acrylamide gels for the second dimension gives more satisfactory results than the gradient gels. This avoids gel-to-gel variation in the acrylamide gradient that inevitably occurs from time to time, even though these are commercially obtained, and thus "*machine-poured.*" Machines are operated by humans, thus it is not shocking that the gradient in one gel in a batch, or more frequently between batches, is not identical with that in another batch. With single percent gels, this gel to gel variation is much less of a factor.
5. Number of biological replicates: when processing multiple samples, it is important that the total number of biological replicates, and whether there will be technical replicates, is decided *before* the analysis by 2D gels. To minimize variation due to either technician error, or day-to-day variation, or instrument variation, all samples are listed, then *subjected to a randomization* by a statistician, so that on any given day, a set of samples is processed in a particular but random order. If technical replicates are being analyzed, the replicates for a given biological sample should be processed on different days, to address day to day variations. If access to a statistician is not logistically feasible, writing the sample numbers on pieces of paper, and literally picking them randomly out of a container, one by one, is MUCH better than nothing.
6. Loading gel samples randomly vs systematically: loading of conventional SDS-PAGE gels with several samples is typically done so that the order of the samples makes experimental sense; the control is usually at one side, and the various treatments in increasing or decreasing order, going away from the control. But, *it is preferable to load the samples randomly*, to avoid a bias introduced by some factor affecting the gel horizontally in a systematic way. This way, if there is an effect of the treatment,

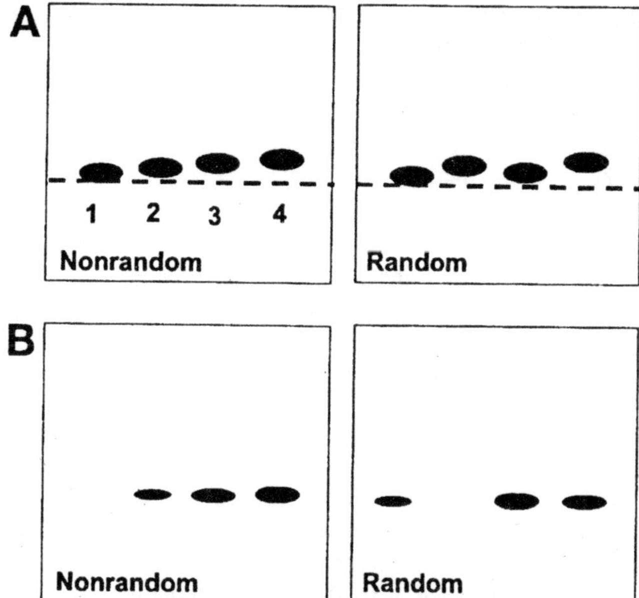

Fig. 15.8. Randomization of samples on sodium dodecyl sulfate (SDS)-polyacrylamide gel electrophoresis (PAGE) gels or Western blots facilitates detection of nonrandom data.

the pattern of signal will be asymmetrically distributed across the different lanes. Densitometry can reveal what the quantitative differences were. The gel lanes can be repositioned *in silico*, to generate a figure that is easier to interpret in a publication.

7. 2D gel staining: protect Sypro Ruby from light! We have found that staining the gels overnight in Sypro Ruby gives the most consistent results. Care must be taken to protect the gels from light during the entire staining process, until their images have been acquired. Once they have been scanned, they should be stored under light-tight conditions, until absolutely sure they are no longer needed.
8. The order of staining makes a difference: gels can be poststained after Sypro Ruby with Coomassie Brilliant Blue, but they cannot be stained with Coomassie Brilliant Blue first, then with Sypro Ruby.
9. Practical considerations in staining multiple gels simultaneously: regardless of what type of stain the gels are being processed with, have enough solution in the dish to allow the gels to float freely off the bottom of the container during the staining/destaining

process. Multiple gels can be stained in the same dish, but be sure and either nick them at the corners asymmetrically so that each is distinguishable from the others, OR, stack them in a particular order in a container in which there is not enough room for them to be re-ordered once they are laid on top of each other. Make sure that there is sufficient liquid in the container that the gels do not "catch" on each other; this will result in tearing of the gels. Finally, make sure that the "*agitation*" is of sufficient speed that the gels do not "sit" on the bottom and stain or destain nonuniformly. When you first put all the gels in their containers on the agitator, watch the agitation. The speed setting may have to be adjusted (increased) once the full weight of all the gels in their containers is on the platform.

10. Stick with Coomassie Blue or Sypro Ruby for MS: Although Coomassie Blue and Sypro Ruby stains are not as sensitive as silver stain, the latter modifies lysines and arginines, whereas Coomassie Blue and Sypro Ruby do not affect proteins covalently. Thus, unless you know the proteins of interest are low abundance, and will be detected with difficulty without silver stain, it is recommended to stain with either Coomassie Blue or Sypro Ruby. Variations of the original silver stain methodology have been described, however, that allow for MS analysis following silver staining. Alternatively, image analysis can be done with the silver-stained gel (although the dynamic range of silver stain is notoriously narrow), and the sample re-run with higher loadings and stained with Coomassie or Sypro Ruby, to enable MS analysis, once the spots of interest are indicated by analysis of the silver stained gels.
11. 2D gel proteomics is an iterative process. Once statistical analysis has identified a list of spots that are significantly different and should be "picked" for MS analysis, this list should be held by the operator in front of the computer, and the gel images checked visually to see if the spots determined to be significant by the statistics are actually different in the images. No one analytical process is perfect; it is possible that the statistical analysis has indicated a nondescript smear as a spot that was significant between groups, whereas to the user it appears as a smear that should be deselected as a spot. Statistical analysis, however, is blinded to the images, and the analysis is usually done with spot intensities normalized to the total spot volumes on a gel, such that zooming

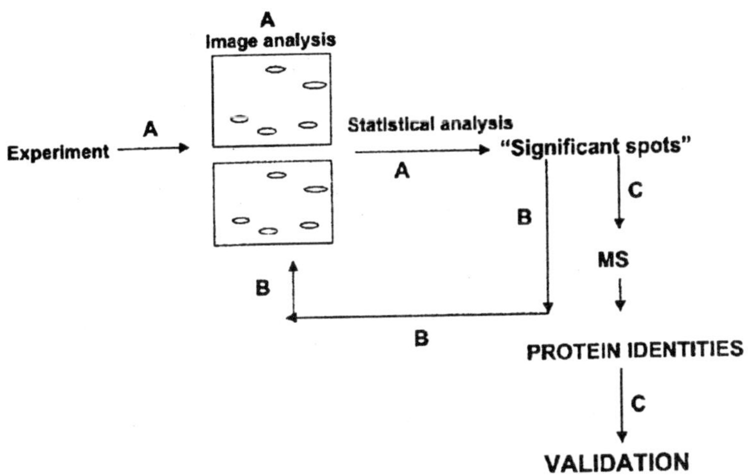

Fig. 15.9. Effective two-dimensional (2D) gel proteomics involves iterative rounds of analysis.

in on a given spot to assess whether it is different in abundance relative to the same spot on another gel may suggest an answer inconsistent with the statistical analysis. But the important point here is that it IS important, once statistical analysis has been done, to go back to the images and assess, spot by spot, whether (1) the spots being compared are the same spots; (2) whether it is a "spot" or a "streak" and, if the latter, whether it should be deselected, or detected differently; and (3) whether the numerical relationship suggested for a spot is visually detectable. Because of the complexity of spots in many 2D gels of biological samples, it is difficult to *see* whether a spot is increased in intensity between two gels unless it is by a factor greater than 2–3. But, overall, one should be able to discern that differences in spot intensities between different groups are consistent with the statistical analysis. If many of the differences though are in the range of 1.5- to 2-fold, these especially should be validated by independent methods, such as Western blots.

12. DIGE: alternative to single sample, single gel: In this chapter, we described the 2D gel protocol in which a single sample is resolved on a single gel, stained with Sypro Ruby or Coomassie Blue, and multiple 2D gel images are analyzed by software assisted image and statistical analysis to determine "significant" spots. It would not be fair to end this chapter without mentioning that we are implementing DIGE using large-format gels to complement

the smaller Criterion 2D gels. Recent experiments involving large sample sets, especially clinical samples, have demonstrated that a particular tissue from a defined group of animals, such as normal female rats of the same age, can have sufficient variability in its proteome that comparison of single samples on single gels requires several biological replicates. In addition, to ensure that a given pattern on a 2D gel is "real" for a sample, technical replicates of each sample may need to be done. Eight animals per experimental group is usually suggested to provide sufficient power for an experiment. So, for a single sample/single gel experiment, if there are 2 dietary groups, 8 animals per group, and 3 gels per animal tissue (liver, for example), this results in 48 2D gels. With these numbers of gels, we have found that there is sufficient variation among animals in a group that measurement of statistically significant *experimentally*-induced differences above biological variation within a group is a nontrivial challenge. For these rationales, we have implemented the DIGE methodology, where similar protein samples to be compared, such as brain homogenates, are labeled covalently with different low-molecular-weight fluorescent dyes, and co-electrophoresed on the same gel. A mixture of ALL of the samples in an experimental set labeled with a third dye is also electrophoresed on the same gel. These dyes, cy3, cy5, and cy2, fluoresce at distinct wavelengths, and are thus quantifiable on the same gel. The beauty of the DIGE technique is that, silly as it sounds, the same protein migrates with itself. Of course it should; it is the same protein. But in fact, as a result of gel-to-gel variation, the same protein can often be found at slightly or distinctly different x,y coordinates from gel to gel. Calculation of the biological variation distinct from the technical variation for a given spot becomes a statistical challenge, especially for low-abundance proteins. With the DIGE technology, although gel-to-gel differences still factor in because more than one pair of samples is usually analyzed, having the mixed standard *on every gel*, and having pairs of samples on every gel greatly reduces the variation that must be dealt with. Recent reports suggested that although the cydyes are expensive, the greatly reduced number of gels and the internal standard enable generation of statistically meaningful data from a clinical study with as few as six gels for a dataset of 18 samples. The virtues of DIGE have been discussed at length recently.

However, there is no perfect technology. DIGE has been suggested anecdotally to involve too much variation in labeling of the proteins. Especially because the basis of the technology is minimal labeling of lysines, as more labeling induces mass changes that make the labeled spot migrate differently from the unlabeled spot, a few % variation—when the desired extent of labeling is only 1–3%—may be too much to tolerate. Moreover, the gels must be scanned immediately after the electrophoresis; a fixation protocol so that they can be left chemically stable until scanned has not been developed, so the running of the second dimension must be timed so that access to the scanner for prolonged periods will be available. But the good news is that the gels *can* be scanned immediately after electrophoresis; they do not require further staining. Furthermore, with the gel chemically affixed to the plate, gel handling, tearing, or uneven swelling are not issues. Also, the DeCyder? (GE Healthcare) image analysis software dedicated to DIGE gels has recently been upgraded to include a module that processes the image analysis data from DeCyder analysis through the standard statistical algorithms including principal component analysis and discriminant analysis, in addition to t- and F-tests (GE). Detailed protocols for implementing cydye labeling of protein samples, the DIGE methodology, and the capabilities of DeCyder can be downloaded from the GE Healthcare website. We anticipate that, for complex proteomes, such as brain, that are especially impacted by aging, using technology such as DIGE to reduce technical variability between samples and enhance resolution of real differences will contribute to understanding the molecular basis of aging-related processes leading to pathology.

13. Meaningful proteomics data, specifically with regard to age-related questions: *"real" data*. Data can be "real," but are they meaningful? This is a central dilemma with data generated by high-throughput methodologies such as 2D gels and MS. Gel spots can be different in abundance between treatment and control, or between senescent and dividing cells, but do these differences *cause* the biological phenomenon being analyzed, or are they "real" but not causal. Thinking simplistically, there are three categories of "real" differences in any proteomics data. The first is those differences in proteins that are involved in the response of the system to the stimulus, or in defining the age-related disease condition. The second is those differences in proteins that may

occur in response to the same stimulus, or that may be correlated with the development of the age-related disease, but in neither case are necessary for the response. Finally, a set of protein differences may be "real" in a disease condition, or in response to a stimulus, but these may be part of the cellular response *to its own response*, an attempt to either maintain homeostasis by blocking the changes in other proteins, or a protective response to an insult such as oxidative stress. All of these categories of protein differences are real and statistically significant. The task for the investigator, once an experiment has been done and reams of proteomic data generated, is to decipher the different categories of protein differences, allowing for the possibilities discussed.

14. A condensation of the guidelines initially presented by Carr et al. (2004) for the publication of MS protein identification data is as follows:
    (a) Authors should describe how peak lists were generated to be used with serarch engines, identify the search engines used, and describe how peptide and protein assignments were made with that software—the scoring thresholds should be identified. The version of the database and the number of protein records in it should be given.
    (b) Information should be given regarding sequence coverage—note that different forms of the same peptide should count only as one peptide, as should Met-ox and deamidated forms.
    (c) When only one peptide is reported to make a protein sequence identification, the *m/z* and charge state of the peptide should be provided as well as the fitted score for the MS/MS spectra.
    (d) Where biological conclusions are being drawn about a peptide matching to a protein (or its posttranslational modification), the MS/MS specrum should be provided (either in the manuscript or in supplementary materials).
    (e) Peptide mass fingerprint data are allowed, but a higher level of stringency is required. Information must be given about mass accuracy and mass resolution, the calibration method, and whether common contaminant peaks have been removed. The coverage of the proteins sequence should be given and the unmatched peaks listed. As for (a), the scoring schemes should be described.
    (f) Identify whether multiple protein records exist for each sequence and the particular sequence chosen was selected.

(g) It is recommended, if not required, that all MS/MS spectra be supplied as supplemental material. This can be accomplished by providing access to websites holding the data.

15. Recognize proteomics data for what it is: it is not the goal of this chapter to discuss in depth the various ways in which proteomic data can and should be followed up or validated; that would require another chapter. But clearly, differences in abundance for a list of specific proteins is the end of one experiment, but the *beginning of another, or several others.* Which protein changes are followed up or validated will depend on individual experience, what is already known about the system, whether one is interested in the evolution of an age-related disease process, or how the system tries to protect against it. Several of the methodologies and experimental paradigms described in other chapters in this volume should prove invaluable in providing a system that can be mined for proteomic data, or that can generate data to complement or validate proteomics data from other experiments.
16. Blue Native gels: investigators have begun discovering the possibilities for the study of protein interactions in age-related disease with the BN methodology. The power of this methodology is that, on one first-dimension gel, protein complexes in multiple samples can be resolved, and assessed in the second dimension as to how they are different in either disease or in response to a stimulus. Recent studies showed, for example, that liver mitochondria undergo early proteomic and genomic changes correlated with early stages of alcohol-induced liver injury. The potential for delineating both qualitative and quantitative differences in complex composition is possible with this 2D electrophoresis method. In conjunction with conventional Western blot analysis, BN electrophoresis offers a readily accessible method for initial analysis of protein associations that can provide powerful initial analysis of protein differences functionally implicated in disease.
17. 2D BN vs conventional 2D gels: the BN and conventional 2D gel procedures resolve proteins in qualitatively different ways; therefore, if there is sufficient quantity of a biological sample, it should ideally be analyzed by BOTH BN and "regular" 2D gels. If very little is known about global protein differences between experimental groups, regular 2D gels are a good starting point. However, it should be kept in mind that unless the sample is fractionated in some way, by either sub- cellular fractionation or

biochemical separation (i.e. by chromatography), the 2D gel analysis will reveal only the "low-hanging fruit," that is, the most abundant proteins that are detected within the 100 or 200 μg loaded on the 2D gel. On a BioRad Criterion gel, the typical resolution will yield about 300 spots resolved from each other. On larger-format gels such as those made available by GE Healthcare, many more spots are resolved and detected (up to several thousand), but similar issues apply in that proteome subfractionation will greatly enhance the analysis even with the larger gels. The value of the BN gels is that functional protein differences that occur in aging-related processes may be readily obtained, if they involve protein–protein interactions. Visual assessment of the initial BN 2D gel may reveal that additional BN gels must be run, with perhaps a different acrylamide gradient for the first dimension, to resolve bands in a particular molecular weight range. The BN gel parameters given here were developed for analysis of mitochondrial complexes, and are not necessarily optimal for analysis of other membrane-associated complexes, or for cytosolic complexes.

Experimental Issues in Aging Research that Impact Proteomics Approaches

The purpose of this chapter was to present 2D gel and MS-related approaches to the study of aging-related processes. Other chapters in this volume deal with multiple other methods. It should be pointed out that, as with many of these other methods, experimental conditions will dictate the success of analysis by proteomic approaches. Because of the cost and labor-intensive aspects of proteomics approaches, particularly if carried out by a facility at which charges are involved, one should have several goals when carrying out proteomics experiments to study aging processes, including reducing proteome complexity and increasing biological specificity. The former can be accomplished, as discussed earlier, by subfractionation in various ways. The biological specificity can be addressed by predetermining the particular age at which the proteomic analysis will detect differences, for example, by behavior studies. If the particular age is not known, then obtaining a proteomic profile of the changes at different time points during the aging process becomes a proteomic experiment. The same titration should be carried out for a cell culture experiment to make sure that the stimulus is employed at a concentration that will maximize the desired phenotypic effect. Sometimes, biological specificity can also be enhanced for a proteomic experiment simply by examining functionally different regions within an organ, for example, the hippocampus within the brain. With proteomic data from aging studies, however, one must keep in mind that differences correlated with aging phenomena are just that—a mixture of protein differences, some of which may cause the aging process, some of which may be "real" differences but not causal, and some of which may be attempts by the system to counter the "aging" process. The proteomic data should be considered as "*snapshots*" of the biology; the more one can get, the more information one has. A single snapshot may be data- rich, but poor in actual information.

INDEX